The

Universal

Urge

Also by J. H. Prince

ANIMALS IN THE NIGHT
ANATOMY AND HISTOLOGY OF THE EYE AND ORBIT
  IN DOMESTIC ANIMALS
COMPARATIVE ANATOMY OF THE EYE
THE RABBIT IN EYE RESEARCH
VISUAL DEVELOPMENT
YOU AND YOUR EYES

The Universal Urge

Courtship and Mating Among Animals

J. H. Prince

THOMAS NELSON INC.
Nashville · Camden · New York

First edition

Library of Congress Cataloging in Publication Data
Prince, Jack Harvey.
 The universal urge. Courtship and mating among animals

 Bibliography: p.
 1. Courtship in animals.   I. Title.
QL761.P73   591.5′6   72–6950
ISBN 0–8407–6234–8

Acknowledgments

Acknowledgment is gratefully offered for opportunities granted by the Director of Taronga Zoo Park, Sydney, Australia, and by the Director of the Australian Museum, Sydney. Many of the illustrations were made in these two educational centers, and Figure 40 was provided by the Director of Taronga Zoo Park. Many members of the staffs of these two institutions were also helpful, and the Fisher Library of the University of Sydney, Australia, provided considerable bibliographical resources.

Other centers used for observation were Melbourne and Adelaide Zoological Gardens; Lone Pine Sanctuary, Queensland; the Australian Reptile Park, Gosford, N.S.W.; the Adelaide Museum (which was an unusually productive source). Figure 12 was made in Perth Zoo, and Figure 45 was made in Adelaide Zoological Gardens, both in Australia. Figures 48 and 49 were made in Columbus Zoo, Ohio.

I am particularly indebted to Mr. Ronald Strahan, the Director of the Taronga Zoo Park in Sydney, for reading the manuscript of this book and making a great many useful suggestions, which have helped to improve it. Apart from this, great effort has been made in the writing to prevent the book from becoming too scientific for the general reader, who is its intended audience.

To
Ayesha,
giver of life

Contents

Preface

There are countless ways in which animals reproduce themselves and thus perpetuate their species. The complex nature of what we have chosen to call "courtship" is of absorbing interest; courtship is not always what it seems to be, because greeting, aggression, and the preludes to mating are often tightly interwoven with each other, and may even be parts of a single emotion.

Everyone knows that there are occasions when man controls his actions intelligently, but with few exceptions we do not ordinarily credit other animals with intelligent thinking. Few people can believe that when a lower animal does something that appears to duplicate a similar action in man, it does it for the same reason as man. And it must be admitted that much scientific investigation supports this skepticism.

The reader is expected, therefore, to be as open-minded about the reproductive habits of all animals as scientists try to be, accepting only that all of them have developed the habits that suit their particular way of life, and that any changes in those habits would be impossible without changes in other habits also, perhaps even of environment and anatomy.

Scientific names have been minimized in the text, but animals that are numbered are included in a glossary at the end of the book where their scientific names are given for those wishing to know them. The common names are included in the Index in the conventional manner.

The

Universal

Urge

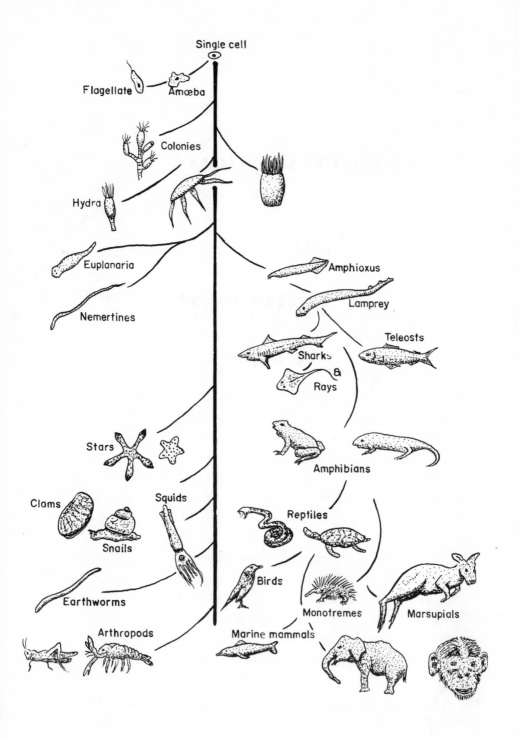

Introduction

The process of reproduction may appear to be quite different at different levels of the animal kingdom, but, in fact, there are always areas of possible comparison—from the lowest organisms to the mighty elephant and whale. Not all animals can be differentiated into two sexes; some of the simplest are able to reproduce asexually. But when those that do separate into sexes are examined, similarities begin to increase, and when one considers animals within one class, such as mammals, there is even more similarity. In fact, the reproductive system of all placental mammals would seem to many people to be identical.

With close observation, however, there is one noticeable area of considerable difference between genera and at times even between closely related species. This is in the activity that leads up to mating and reproduction, or, as we call it, the "pattern of courtship."

Descriptions of all the methods of reproduction can be found in detail in biology textbooks. Anyone interested in animal reproduction needs a dissecting laboratory and a microscope to follow these books in a practical way. But for the study or observation of courtship, patience and perception are the only essentials. And one must add, of course, time.

The approximate positions of some of the classes and orders of animals in the evolutionary scale.

THE SEARCH FOR A MATE

Before courtship can begin animals must find each other. Those that live in integrated groups have no problems in finding a mate; some are always at hand within the group, and they merely have to be persuaded to accept the attentions of the opposite sex. Those that gather in communal meeting places in spring, such as lakes or ponds, and those that migrate in large numbers to specific nesting sites or areas have no great problem either. But for all animals that either live solitary lives or remain in segregated herds or flocks, the search can sometimes be long, and encounters can even be purely by chance. This is especially so for those living in the lightless depths of the oceans.

Many species of animals have few superficial features that identify their sex, so individuals must recognize each other by a characteristic odor or a signal that reveals their sex. In some fish and birds the sexes are colored differently in the breeding season, are different in shape or size, or give specific signals by movement, and these animals can approach each other with certainty. This is true of the fireflies, which have specific flash patterns, and the deep-sea creatures with characteristic patterns of body lights.

Finding a mate visually through the medium of flashing lights or color, however, is obviously restrictive in scope and may entail considerable searching. Vocal calling, which in some circumstances may carry for great distances, enlarges the area of contact. It makes identification possible twenty-four hours a day, so this method is used by a number of animals—fish, amphibians, reptiles, birds, and mammals. It does not matter if only one sex calls, and the other does not answer so long as the latter is attracted to the calling and goes toward it. Frequently, both sexes do vocalize, and this shortens the searching time because both move toward each other. Even more animals make sounds during actual courtship.

Vision and hearing are both used for species and sex recognition, and the sense of smell is used to confirm the identity once a pair has been brought together. Some animals use odor extensively for searching as well as for identification. Snakes use an extremely sensitive organ in the roof of their mouth for tracking down mates through odor, and we know that some fish and amphibians recognize each other by odor. In some fish that live at the bottom of the sea, the female releases a chemical signal into the water when she is ready to spawn, and the male detects its odor to find her.

Many mammals also use their excellent sense of smell to investigate at close quarters and to establish the identity of any animal encountered, as well as to search. In those animals capable of following a scent, that ability is certainly more useful than vision.

Just as certain flies and fish carry or flash identifying light patterns, some fish living in the lightless depths transmit specific patterns of electrical currents. These currents are like signature tunes, which may be the result of the pattern of an individual's particular muscle movements. Anyway, these currents identify the animal to its own species, and the sexes differ enough for this to be a means of finding each other in the dark. Once again the range of the current is very limited, and therefore it is only effective for identification purposes once a member of the species has moved to within that range. This electrical transmission is quite different from that used for "shocking" by the so-called electric fish found in fresh water.

COURTSHIP

When male and female animals of the same species find each other, this does not always ensure that mating will be accomplished. In many species the male must also win the female's acquiescence or dominate her. He may go through what sometimes amounts to a long, ritualized courtship. Many primates rely much more on readiness to mate than on courtship. They may enjoy a communal sharing of favors, or a leader may own a number of females, which he takes by sheer physical strength or dominance of personality. Man has developed quite an elaborate courtship in some cultures. Birds are among the most interesting courtiers, for many of them carry on a continuous ritual throughout the periods of mating, incubation of the eggs, and even rearing the young.

Courtship is much more complex than is generally supposed. As a prelude to mating, it is often tightly interwoven with both greeting and aggression, and at times, aggression against rivals which wander into marked territory and courtship greeting are almost indistinguishable. Then, too, other factors, such as environment, climate, food needs, and evolutionary development, all play some part in forming the actual courtship procedure, its duration, the time of the year it occurs, and whether or not it is monogamous or polygamous, in isolation or in groups.

In some animals these factors are also important in the raising of offspring—whether by the mother alone or by both parents, whether in a nest, a burrow, or within a free-ranging group. Some of these factors may even be important in determining whether the species bears live young or lays eggs, whether there are many or few newborn, and whether they are large or small compared to their parents.

Members of different animal groups and often members of a single group live in vastly different habitats, and animals vary considerably in size. Mammals, for instance, vary from an inch or fourteen feet in height to nearly eighty feet in length, as in the blue whale. Quite obviously, these variations must influence courtship activity.

A few mammals and a great many birds have very elaborate courtships involving dancing and display. The birds are easier to observe, since their activities usually take place in daylight and often in open areas, whereas the courting activities of many mammals are frequently confined to the night or to secluded places.

Some birds and mammals have something in common that is seldom seen among lower animals, apart from a few fish. This is a period devoted by the mother, in rare instances by the father, and sometimes by both parents to preparing the young for adult life. There may be no actual teaching of the young in the sense that humans teach their young, but they are protected and given examples, which they undoubtedly follow and learn from.

There is quite a similarity between certain forms of animal courtship and those acted out by some human groups. In their mating, humans are involved with such things as displays of possessions, which *might* possibly have its equivalent among animals in the amount of territory controlled. Man's economic or environmental conditions *might* be like an animal's access to food and water. Vocal persuasion in man certainly seems to have its counterpart in the intimate sounds made by some animals. Tradition and habit are well rooted in both man and animals. Social prospects in man are perhaps like an animal's nesting site or hunting area. Beauty of color and form, and personal charm and attentiveness to the opposite sex, can be found in countless species, but whether or not the last is motivated in animals in the same way that it is in man is questioned by some experts.

Animals may show mating responses to color display, to singing or calling, to sheer physical strength, or to combinations of these and other things such as odor. But above all, animals respond to the fact that their physiological condition is right for mating. The specialized displays and courtship activities ensure that mating remains within the species, avoiding hybridization and changes in species forms.

THE HORMONES

The onset of courtship behavior as well as actual sexual activity is brought about by the action of glandular secretions known as hormones. These appear to be essential in all vertebrates and all but the simplest invertebrates. Glands that are involved in this include the pineal, the pituitary, the thyroids, the adrenals, and, of course, the gonads. The sex-stimulating hormones are of two kinds. Gonadotropins are secreted by the pineal and pituitary glands and, as their name implies, act on the gonads, inducing them to produce their own vital gonadal hormones (the second kind), which are then released into the bloodstream.

Of the latter, testersterone, which is secreted by the male testis, is responsible for the seasonal return of the sex function in spring, puberty changes, the growth and development of the whole male reproductive tract, male sex color, scent glands, antlers, mating behavior, aggressive defense of territory, bird song, and so on. This action is so positive that if the females of many species are injected with testersterone, they will lose their female characteristics and assume those of males.

This effect can be induced in another way. The female peafowl will grow male plumage, becoming the strutting ornamental peacock, if she is deprived of the female hormone estrogen through removal of her ovaries. So specific are these hormones that if the ovaries are removed from female birds, causing them to take on male plumage, and then they are injected with the female sex hormone, the female plumage is restored. A female canary that is injected with the male hormone will sing like a male, and the same effect is seen in domestic chickens. These hormones also seem to play some part in stimulating the migration urge in birds.

The master switching system for the whole vertebrate sex hormone complex and probably for all vertebrate gonadal development is the pituitary gland. This gland triggers the growth changes that occur in the sexual systems of both males and females before these become fully functional. One of the pituitary hormones, prolactin, plays a part in the production of milk in mammals.

In most nontropical animals, the pituitary gland seems to react to the length of the day. It can also be stimulated by the mere presence of other members of the species, and in some females by the mere sight of a member of the opposite sex. Such reactions are obviously likely to lead to courtship activity.

This is only a very small part of the hormone story, of course. It is enough, however, to show the complexity of this reproductive chain, in which light intensity or the length of the day can stimulate a gland which, in turn, activates courtship behavior, ovary and sperm ripening, and eventual mating and conception. Environmental conditions and climate can inhibit breeding by acting through the glands that stimulate the breeding cycle or by depriving animals of suitable nesting and rearing sites. Conversely, they can also stimulate glandular action and thus all the facets of the breeding cycle. Thus, environment can influence more than safe and appropriate courtship and breeding. It can influence whether or not these actually occur.

Part I

Mating
Behavior

1. The universal urge in fish, amphibians, and reptiles

FISH

Fish have a host of courtship patterns in which one sex pursues the other or tries to woo the other in some special way, such as with color display, acrobatic encirclement, fin and tail extension, and even sounds. Sometimes nest building is part of the courtship ritual, although not all species build nests. Some carry their eggs, and even the newly hatched fish, in their mouths; and others have a special pouch for brooding.

Shoaling

Not all fish display these courtship patterns, of course. Many merely swim in shoals and spawn haphazardly and in unison without any form of display, the main stimulation being the close proximity of the two sexes. The habits of most of these fish are not known. Some of them, such as the striped mullet, seem to form tight groups within the shoal, each group including a number of males, which surround a female and take turns spawning her.

Nest-Building Species

Many species, however, such as the blennies, are noted for complex courtship and mating behavior and for the care of their eggs and newly hatched fish. This greater use of ceremony and ritual may be correlated with the production of fewer eggs than other fish produce, or in some cases, with eggs of greater individual size in smaller numbers.

One fish that definitely pairs up is the salmon. The males, exhausted from their long migration, fight for the spawning sites, while the females prepare pits for their eggs. When a female picks her mate, she takes the strongest male she encounters. If he is driven away by others, she merely makes another choice. Once they have copulated, the female moves a little upstream to hollow out another nest. The gravel she removes from this drifts down with the current to cover the eggs in the first nest. This is repeated until all her eggs are laid.

The male salmon changes to exotic coloring when in breeding condition. This change not only attracts females, but, perhaps just as important, it warns rivals. Some male salmon change their body shape also. Others merely change the shape of their fins, and in some species the males' head shape changes as they move toward their spawning streams.

The male of the European stickleback [2] (and other species) leaves his shoal in the spring, changes color to his breeding coat, his underside becoming flushed to a brilliant crimson, and takes up an individual territory, where he builds a nest. So strongly is this color change embedded in the breeding pattern of the species that a female will respond to a simple lure if it is colored like the breeding male. The male is stimulated by the shape of the female's belly and courts the female with the thickest belly. Similarly, he will respond to a suitably curved lure.

The stickleback builds his nest with plant material, cementing it together with a sticky secretion. When this is done, he shows the nest off to passing females, enticing them into it to lay their eggs, which he then fertilizes. He may do this with a number of females, then he drives them away and patiently fans the eggs with his fins until they hatch.

In the tropics particularly, and especially in tropical rivers, we see quite impressive fish courtship displays. Few people have not seen or heard of the displays put on by some of the tropical aquarium fish such as the paradise fish [3] or the Siamese fighting fish.[4] The male of these and other related species displays by spreading his tail and fins, and quivering his way around the female in a kind of floating dance. His color is heightened by the release of hormones into his bloodstream. He builds a nest of air bubbles and mucus at the surface of the water, and after he has spawned the female, the fertilized eggs rise into the bubble nest. If any eggs drift downward, he picks them up and takes them back to the nest.

When the female is empty of eggs, the male drives her away and then guards the nest continuously. If any eggs or young fish fall out of the nest, he takes them into his mouth and blows them back up again. However, as soon as the fry, or young fish, reach a certain stage of growth, the male parent turns cannibal, if they have not dispersed into hiding.

Mouth Brooding

Several species, instead of taking eggs and fry in the mouth and returning them to the nest, retain them in the mouth. They are even called "mouthbreeders," although mouth brooders would be a more accurate term.

In *Tilapia macrocephala,* a mouth-brooding species found in Africa, both male and female participate in the courtship display, which consists of slapping the tail, quivering the body, and expanding the throat area. Both sexes also indulge in a considerable amount of nibbling at each other, which seems to stimulate them to spawn, although strangely this behavior increases in the female after she has been spawned.

Before being spawned, the female goes through a nest-building routine, digging a hollow in the sand, but no more than

1. In this mouth-brooding species the male incubates the eggs in his mouth, and the young take refuge there for a while after hatching. Notice how large his mouth is compared with that of the female behind him.

a minute after the eggs are laid, the male begins to pick them up and hold them in his mouth. On the rare occasions when the female helps with this, she begins to pick them up much later than he does. The eggs hatch in about five days, and the male holds the fry for another seven or eight days before turning cannibal. So for at least twelve days he does not eat at all.

There are more than forty different species of mouth-brooding catfish in which the male carries the eggs, all the while going without eating.

In the Egyptian mouthbrooder [5] (mouthbreeder), it is the female that takes up and protects up to fifty eggs in her mouth. Once these are hatched, the fry swim in and out of her mouth, returning for protection. Once they reach a certain size, however, their parent, as in so many species, turns cannibal.

The number of eggs a fish can retain this way naturally depends on their size. The male *Apogon imberbis*, a cardinal fish in the Mediterranean Sea, carries about 20,000 eggs, each a fiftieth of an inch in diameter. A related species in Australian waters, *Apogon conspersus*, carries only 150 eggs, but they are a sixth of an inch in diameter.

Pouch Brooding

The quaint little sea horse [6] seems to have reversed the normal patterns of parenthood and breeding in more ways than one. Not only is the female the one that does the courting, but the male incubates the eggs and the young in a pouch developed especially for the purpose, which some observers have likened to an external uterus. When the female has interested her male, she grasps him with her tail, and they swim closely locked together, swaying as they go until they reach the surface. At that point the male presents a pouch on his abdomen into which the female extends an ovipositor, a brightly colored papillary projection. Through this she discharges her eggs into the male's pouch, which stimulates him to release his sperms among them.

Some male pipefish, relatives of the sea horse, also have this pouch; others merely carry the eggs attached to their abdomens. The sea dragon,[7] another relative of the sea horse, has a similar but not identical arrangement. The male incubates the eggs within a spongy matrix low on the abdominal area.

Other Mating Habits

There are other mating habits among fish that may seem strange, but all have been adopted for their contribution to successful spawning. Bitterlings [8] lay and fertilize their eggs in the gill chambers of mussels, the female projecting a long ovipositor to inject the eggs and the male discharging his sperm into the gill chambers. Here the eggs develop and hatch. They are protected inside the shell, and the young escape when they hatch.

Grunions ride waves onto the beaches on the Pacific coast of the United States; the females stick their tails into the wet sand and drill down to lay their eggs, while the males curl around them and discharge their milt to fertilize the eggs, before riding the next wave back to the sea. Two weeks later a high tide washes the eggs out as they hatch. Several other species of fish appear to emulate the grunion.

The female *Aspredo,* a South American catfish, does not trust her eggs to any care but her own. She is spawned at the bottom of a river, and immediately rolls on the eggs until they have stuck to her abdomen. A skin then grows over and encases each one. That skin also attaches each egg to her body with a stalk of tissue through which blood vessels pass to nourish the developing embryos in much the same way that the embryos of higher animals develop with the aid of a placenta.

Parasitic Mates

Some of the deep-sea anglerfish defeat the need to search for the opposite sex in the dark. Once they find each other, they never part again until they die. The male is very small, often no more than 5 percent of the female's size, and once he finds her through odor tracking, he attaches himself firmly, usually near her genital orifice, and then he gradually grows into her side, giving up his independent feeding and circulation, just fusing with her completely. All his organs degenerate except the gonads, and these develop excessively, so that he becomes nothing more than an attached male sexual organ, sometimes no more than a thousandth of the weight of the female. Females with several males fused into them in this way have actually been caught many times.

AMPHIBIANS

Certainly, the calling of "love-sick" frogs and toads is known to everyone who lives near water. Many male amphibians have inflatable vocal pouches or sacs, and when they are ready to mate, they sit in ponds or any shallow water and call constantly to attract the silent females to them. Perhaps the sexes recognize each other by the sounds the males make and the very silence of the females.

There are great differences between the calls of various species, almost like the language differences in human groups, and some of these animals are quite versatile, showing a wide range of notes, which they deliver at different speeds. All notes have relatively low frequencies, so they are audible to human ears. One toad, *Bufo valliceps*, appears to utter up to thirty-eight notes a second.

The female frog may be silent, but she has recognition features of another kind. She develops a series of granulations on her thighs, which can be seen or felt, and these advertise her sex to any male as she brushes past him or he past her in the water.

Amphibians use several methods of rubbing the opposite sex with the head in courtship, perhaps what we might call "nuzzling"; and many, such as the newts, use gland secretions to attract by odor, but these are usually confined to the males for attracting the females.

In general, tailless amphibians breed communally, and most have definite mating calls. Some species even seem to produce choruses, with the different notes being uttered in a certain sequence. An American tree frog, *Hyla crucifer*, sings in trios. Usually, mating is accomplished by the male climbing onto the female's back and assisting in the extrusion of her eggs, her movements in turn stimulating him to release his sperm at the same time.

Not all amphibian courtship is in or adjacent to water. That of the *Salamandra* and Gymnophiona is on land, and some frogs also court and lay their eggs no nearer to water than the distance that the young can travel when they hatch out, almost as though they would like to do without water but haven't yet discovered how.

In primitive American salamanders [9] and cryptobranchids, the male takes the eggs from the female when she is trying to fasten them to a stone and rubs his cloaca over them.

Some salamanders do not lay their eggs until the young are either partly or fully developed within them. Two of them, *Salamandra salamandra and Salamandra atra*, retain the larvae, which obtain their oxygen from the mother's oviducts through their external gills. The former species delivers larvae, whereas the latter, and a number of other species, deliver fully metamorphosed young.

The axolotl [10] is a salamander that breeds in the larval stage. The male rubs himself on the female, and the sperms embedded in a mass of secretion from his cloacal glands (still called a "spermatophore") are deposited under her. She takes this into her vent, where the sperms are stored in a special chamber until her eggs are laid and need to be fertilized several days later.

REPTILES

Although reptiles do not have particularly elaborate courtships, some of the features of their behavior are very similar to what we see in certain birds—bobbing, circling, and marking of territory. There is, however, much less concern for the eggs and the offspring than is shown by birds.

Snakes would seem to be the least equipped animals for indulging in courtship. Yet a few of them have a mutual dancelike ritual in which they rear up face to face, intertwine with each other, and copulate. The crotalids and vipers, which are active at night because of their heat-sensitive organs, are among those that perform this ritual. Because of their nocturnal habits, however, it is seldom seen.

Nuptial procedures can occasionally be seen in the colubrids and the elapids. Often the male rubs his lower jaw along the back of the female to stimulate her, probably stimulating himself at the same time; and the males of the natrix group—grass snakes and some water snakes—do this too.

When snakes copulate, the male usually throws coils around the female and brings their cloacas into adjacent positions, but sometimes they just lie close without the gripping coils, merely using a branch or some other object on which to obtain a grip to hold them close together. There seems to be very little competition between males in the breeding season, and apart from elapids, few fight. Most are more aggressive in the breeding season, however.

Sea snakes all come onto land to produce their young. However, they always mate at sea, and little is known of their actual breeding habits.

Land snakes use odor to detect the opposite sex. This is virtually imperative among the *Typhlops,* or blind snakes, and in all of them it is accomplished through a fantastically sensitive organ in the roof of the mouth called Jacobson's organ. The snake's forked tongue is flicked into this organ after it has been projected into the air to pick up odors so weak that no nose would be sensitive enough to detect them. With no true ears, a snake's information about the world around it and the presence of other animals is obtained primarily by way of this organ.

Everyone who has been in alligator or crocodile country at night knows that the males bellow their love calls unabashedly in the mating season, and fight off any rivals that happen along. The male marks the limits of his territory with secretion from

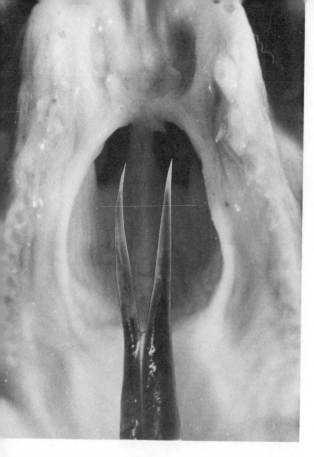

2. In the roof of a snake's mouth is a double pit which is exceedingly sensitive to odor and perhaps to taste. The snake's forked tongue flicks out to pick up scents and then flicks back into these pits known as Jacobson's organ.

his musk glands, and this not only attracts females into the area but also warns other males.

The male and female swim together, gradually increasing their speed; and the male rubs his throat against her snout, marking her with his odor, and vocalizing freely before he copulates with her.

Turtles that live in the ocean have probably never been observed, and nothing is therefore known of their courtship. The female comes ashore at night on lonely islands to lay large numbers of already fertilized eggs in a hole that she digs laboriously in the sand, leaving them at once after covering them, and returning to the sea.

Tortoises are easier to observe; but some of them also mate in water, and only a limited amount is known of their courtship rituals. The male tortoise *Gopherus polyphemus* circles around the female, stopping every now and then to bob his head toward her, and this apparently attracts her. She may nod too. He moves in to her and bites her forelegs, her head, and the leading edge of her shell. If she is suitably impressed, she goes into reverse for half a circle, then, when she stops, she stretches out her hind legs and turns so that her hind end is near his head. She protrudes it from her shell, and this is his cue that all is well; he mounts her and they copulate.

Certain species of tortoises make sounds when they approach each other and as they copulate.

Box turtles can close both ends of their shells with hinged lids, so a male must knock on a female's shell until she opens these lids. He then climbs onto her back, hooking the large claws on his hind feet onto the rear of her breastplate, and the rear lid automatically closes on them. He bites at her head and scratches with his foreclaws until she reopens it and copulation becomes possible.

Although the courtship play of most lizards is not easy to observe, the male Galápagos marine iguana [11] makes himself and his intentions very obvious when he blows spray through his nostrils to impress a female. Some male lizards approach the female with what can only be described as gestures of aggression, nodding his head in a manner which, like color in fish and birds, is used to drive away rival males. And, like fish and birds, he deliberately displays any special coloring he may have. Sometimes this means opening his mouth wide, or it may necessitate rising on his hind legs. If this does not drive away a potential rival, he will attack. Then he follows the female, getting ever closer until he can grab her with his mouth, and they copulate.

2. The universal urge in birds

Fish courtship habits may be varied and colorful, but they hardly compare with those of birds, which have brought vocal and sign language to a level of sophistication enjoyed by no other class of animal, even though some of the higher mammals do have some complex patterns of communication. Birds employ sound, color, dancing, aerobatics, parading, gift offering, the building of decorative bowers, even the clearing of dancing grounds. Sometimes they indulge in several of these activities, and most species not only build a nest or some kind of breeding home, but also exhibit considerable care of the young.

Each species of bird adheres rigidly to its own pattern of courtship and breeding behavior. Most breed in solitary pairs, but some breed in groups of pairs or even in vast colonies of several thousands of pairs, creating a noise and confusion that would be impossible for almost any other kind of animal or even bird to tolerate.

MARKING TERRITORY

In almost all bird species, the first step in the life cycle begins when segregated winter flocks scatter, or lone wanderers assemble in groups. A male stakes out a piece of territory from which he tries to exclude every other male of the same species. This territory can be an acre or so in extent, or, if the species breeds in tightly packed colonies, it may be no more than two or three square feet. Whatever its size, this piece of territory is defended vigorously by the one male, and into it he must lure one or more females.

The notes that male songbirds trill in what appears to be such a happy vein are really advertising the individual's species and his whereabouts, and warning all other males to keep away from the area. Often a male goes from point to point within the territory, marking its boundaries clearly by singing from each point, and the denser the growth of the vegetation, the louder the song is. Usually, all other males respect his claim, and mortal combat for occupied territory is rare.

Females seldom sing, although a few may do so in the autumn, but the male's warning to other males probably attracts a female first, because the most aggressive male is the one that also promises the greatest security. In some instances the male bird becomes less assertive as he becomes better acquainted with his bride and even defers nervously to her, but at the same time he becomes more aggressive toward potential rivals. This is seen in the European chaffinch.[12]

Some gestures as well as song are used for both sexual greeting and aggression. The spoonbill raises his crest when he is courting, and he also raises it as a warning to other males. Not all song appears to be aggressive. Those birds that display before a female by aerobatic flying frequently sing at the same time. One might get the impression that the song is then only for the female's benefit, but undoubtedly it is no less aggressive to other males in the area than that of any bird perched in a tree.

When male and female are indistinguishable from each other, the male that holds territory will attack any female that enters it, just as he will attack any male; but because her response to attack is different from that of a male, he quickly changes his attitude.

Having lured a female into his area, a male makes every effort to display his agility, beauty, and virility. He flits around her, showing off any special colors or plumes he may have, fluffs up his feathers to give himself greater bulk, makes all kinds of attractive sounds, and finally, when she has been stimulated to the point where she has eggs ready for fertilization, he asserts himself with copulation.

In many species copulation usually takes place in privacy, perhaps partly because the act always seems to arouse other birds and invite interference. In species that nest in colonies, however, example stimulates copying rather than interference, and anyway privacy would be difficult to achieve when personal territory is only about a square yard in extent.

THE PAIR BOND

Male displays, or at least some parts of the ritual, sometimes persist right through the breeding season, and this serves to hold the two birds together, strengthening what is known as the "pair bond." This bond, like marriage, creates a situation in which one of the partners can go off on food-hunting expeditions without the other deserting the nest or the young before they are reared. After that, the pair bond may break up quite amicably, but it does ensure the bringing up of a new generation to adulthood. Mutual greeting ceremonies also perpetuate the pair bond, and are indeed part of courtship activity. Both males and females of some species indulge in bill clapping, scraping, bowing, and wing spreading when they greet or recognize each other or when they switch positions on the nest.

FEMALE DOMINANCE

In a few species the female bird adopts the courtship role, and there may be special reasons for this. The red-necked phalarope [13] female defends territory and makes approaches to the male. She starts building a number of nests around the area, and finally decides which one will be used.

The emu is another rare and still different example. When the female courts, she is polygamous, and as soon as she has copulated with one male, she goes off looking for another, leaving each one to incubate the eggs and raise the young in the nest he built for her. There may be an advantage in this arrangement in that a larger number of chicks can be raised under adverse conditions.

The female of the painted quail [14] is also dominant over the male, and may even have a harem of several males scattered about. She displays to each male, scratching the ground, calling, and even offering him food. He accepts the task of incubating the eggs and rearing the young.

Now and again a female bird will reverse the role of dominance in copulation too. Ordinarily the male jumps onto the female's back and brings his cloaca into contact with hers. Sperms are deposited and climb to meet her ova, which are on their way into the oviduct, but it is not unknown for a female to mount a male occasionally. Each having a cloaca probably makes this possible, and the union may even be crowned with success.

POLYGAMY

The emu is only one of a number of polygamous bird species. Without any true courtship, male mallards chase and mount any female passing by; but they stay with one chosen female for the rest of the time. The male does not protect his personal female, but he does stay around while other males copulate with her so that he will not lose her entirely. Sometimes a female takes it into her head to resist the attentions of a passing male, and may even do so to the point of death by exhaustion or drowning.

Swamp sparrows, *Melospiza georgiana*, in eastern Massachusetts exercise male bigamy. Each female has a subterritory; the male has favorite singing perches between the two subterritories and visits each female in turn.

BOWER BUILDING AND GIFT OFFERINGS

When male birds of paradise hang their heads down with feathers and wings spread, they have the most exotic courtship plumes and displays of all birds; but their relatives, the bowerbirds, are in some ways more interesting, for they build and decorate special display areas quite apart from their nesting sites. The male builds the bower in which he courts, and in some species in which he copulates with the female, but the female builds the nest in which she lays her eggs and then raises the young without any male assistance.

3. The male satin bowerbird, *Ptilonorhynchus violaceous*, stands in front of his bower, while a female inspects it from beyond.

One exception to this pattern is the regent bowerbird.[15] The female makes the bower, and then after she has attracted the male and been fertilized, she scatters the bower and builds a nest.

There are about nineteen species of bowerbirds, and all are in Australia and New Guinea. Their bowers vary tremendously, from a parallel pair of twig walls to most elaboratley built and decorated avenues around which the males dance and sing.

The decorations may consist of almost anything—pieces of paper, plastic, spoons, shells, bones, stones, berries, flowers, feathers, pieces of glass, dead beetles, anything that may be found within the area, especially if it is colored. Some species have distinct preferences. The blue satin bowerbird, *Ptilonorhynchus violaceus,* for instance, collects mostly blue ornaments, and he rearranges these continually, often picking one up to offer to the female. The great bowerbird,[16] which builds one of the largest of the bowers, sometimes twenty-four inches long and fifteen inches high, shows a distinct preference for red and white ornaments.

Surprisingly, the smallest bowerbird builds the largest and most elaborate bower, and one that is quite different from all the others. This is the golden bowerbird [17] of the highland rain forests. The male finds a site where there are two saplings fifty or sixty inches apart, and builds a pyramid of sticks around the base of each of them. One of these pyramids may be up to fifty inches high, and the other up to thirty inches, but usually they are somewhat less. Although this is quite a feat for such a small bird, it hardly compares with that of joining the two with a long stick, which he mounts into them. He uses this stick as a display perch and decorates it with berries, lichens, and flowers, usually yellow, leaving just a few inches clear in the middle, where he perches.

It is within his bower or its immediate vicinity that a male bowerbird makes his deliberate display, dancing, whistling, or calling. The greater spotted bowerbird [18] raises a crest when a female has been enticed into his bower, then he selects one of his ornamental objects to offer her and begins a dance which terminates when he mounts and copulates with her in the bower.

This offering of ornamental objects, or even just objects, is seen in the most unexpected species. Certain male penguins offer a chosen female a stone or a piece of snow. The sarus crane [19] stretches back his neck and throws his female a piece of straw. The male European heron, *Ardea cinerea*, presents a female with a stick while raising his crest. If she accepts it and raises her own crest, he then collects large numbers of sticks to build a nest. Such gifts can take many forms—grubs, fish, spiders, flies, grass, waterweed, and so on.

The gardener bowerbird [20] clears a large area of forest, maybe ten feet by ten feet, and plants an upright stick in the center, adding other sticks to form branches like a tree. When a female has been enticed into the area, the two of them chase around this avian maypole until they are stimulated enough to copulate.

COMMUNAL DISPLAY

Some birds display in groups, with all the males and females in a flock participating together, the females ultimately choosing from among the males. The elegantly dancing brolga crane is an example. The black duck [21] also has a highly ritualized display, which is performed as a group activity. Quite often, however, black ducks will begin this ceremony, both in and out of the breeding season, only to lose interest and not complete it. This seems to be a duck characteristic, for it applies to other ducks too, such as the chestnut teal [22] and the gray teal,[23] which both display very much as the black duck does. It applies also to the

4. Courting brolgas, *Grus rubicunda*, have quite a dance festival at mating time.

white-eyed duck [24] and the musk duck,[25] which have entirely different rituals.

Sage grouse [26] display on appointed strutting grounds, and there may be as many as a dozen of these grounds in a single valley, where up to five hundred birds will strut and display in spring. Sometimes a hundred birds will gather in one spot, arriving at dawn, the males strutting around with tails raised and feathers spread, so that the plumes on their neck form a lyre-shaped halo behind the head. Two large yellow air sacs on the throat are inflated and bounce on their chest as they strut. They also make a swishing sound by rubbing the wing primaries across their breast feathers. All this is repeated every five or six seconds when courtship is at its peak, and it makes quite an exhibition.

WATTLES AND POUCHES

Many male birds develop special naked decorative wattles or pouches, and some, like the bustard, *Oris tarda tarda*, grow a

5. The male great bustard, *Otis tarda tarda*, inflates a huge feather-covered pouch to boom his warning to rival males when he is courting.

huge feather-covered pouch, which is inflated when he booms his deep warning to other males. Many other birds have throat pouches. The prairie hen [27] behaves just as the sage grouse does, strutting and calling, and inflating its throat pouches. The frigate bird [28] has a scarlet pouch, which he inflates while drooping his wings, calling, and clattering his beak at the female as she approaches. The bare-necked umbrella bird [29] also has a scarlet pouch.

The male ostrich [30] displays by flapping his wing plumage and producing a drumming noise by banging his head against his body. The three-wattled bellbird [31] inflates three enormously elongated spiked wattles at the base of his beak to impress a female.

Exotic Pirouettes

Although the lyrebird, *Menura novae hollandiae,* is not gregarious like the sage grouse, it has one of the most impressive and highly publicized displays of the bird world. The male stakes out a territory of up to three or four acres for himself, and when a female appears, this drab brown bird becomes transformed. Shaking his tail and uttering a whirring call, he scratches energetically at his mound. Then he raises his tail and spreads it over his head like a fan, sometimes uttering the calls of all the other birds he has heard in the forest, mixing them with his own.

The male peacock turns his back on a female to find out if his display impresses her, although he takes care not to put her to this test until he has thoroughly dazzled her. If she is impressed, she moves around to his front. After one or two tests like this, he is convinced she will accept him, and he mounts her.

6. The male lyrebird, *Menura novaehollandiae,* has one of the most beautiful courtship displays in the whole animal world. He spreads his otherwise very ordinary tail into a beautiful harp-shaped halo over his head when he faces his female choice.

7. The male peacock, *Pavo cristatus*, corners a female with his fan-tail spread, and when he has done this for a while, he turns around and presents his rear view.

8. If he has impressed the female, she will move around him to see his front display again. Then he knows he has won her.

Many birds of prey and some hummingbirds display by flight acrobatics, which often involve great diving speeds and exhibitions of control; and in some species of birds the female teases the male to chase her in flight, just like an adolescent girl with boys on a beach.

Although display is usually exclusive to male birds, females of some species do join in. Male and female wood swallows[32] settle side by side on a tree branch or a fence at varying distances from each other, according to the species. One of them slowly rotates with its tail fanned out and its wings spread, and soon the other follows suit, the two stimulating each other. Their wing movements become ever more rapid until finally they come together and copulate. Egrets[33, 34] are indistinguishable by their plumage so far as their sex is concerned, and they also display together, raising their plumes and greeting each other with a short ritual, even when they meet at the nest.

The crested grebes[35, 36] and the red-throated diver[37] are also species in which the sexes display together. Pairs race side by side

9. A pair of crested grebes will race across the surface of the water in unison with their necks arched in one of the most fascinating courtship displays to be seen.

on the surface of the water, their bodies incredibly upright and their heads stretched forward on arched necks. Sometimes they run on the water like this, each holding one end of a piece of weed, which the male has brought up from a dive. The female invites this by diving herself for a piece of weed at the beginning of the display. Frequently, the performance varies considerably.

Some birds do not continue to the completion of their ritual display until after copulation. The red-breasted merganser [38] is an example. First the female lies prone, and the male circles around her, neck outstretched, going through a number of movements that simulate preening and drinking, after which he mounts her and copulates. Then the pair rotate, and the male continues his courtship display, which leads up to both of them making simulated bathing movements.

GROUP BREEDING

Most birds that gather into vast breeding colonies do so on islands or inaccessible shores or cliffs. These are the seabirds—those that wander over vast tracts of ocean for most of the year or migrate great distances from one hemisphere to the other. In spite of their crowding, some of them have notable courtship displays, often both sexes being equally involved. This crowding is believed to be stimulating, one couple's behavior encouraging others to compete with them, and evidently promoting early egg laying.

Although there may be safety in numbers for some birds, another, more negative, factor is also involved. Birds that nest in colonies defend only a very small piece of territory around the nest—just a reaching distance—and when this becomes reduced by the sheer force of numbers, it is very difficult to expand it again. Overcrowding like this provokes the males, so that they fight much more than usual, and causes the females' sexual cycles to be cut short, so it is quite detrimental to family life.

When courtship display, or at least some part of it, such as greeting and recognition, continues throughout the entire period that a pair of birds remains together, it may be very important in minimizing aggression. Such activity is seen particularly in groups of seabirds that collect in hundreds of thousands to breed. Many of these have been well observed and described. One of the recognition ceremonies performed in the courting of mated pairs is the clattering together of their beaks, often accompanied by sounds which must have initiated the expression "billing and cooing." Storks clatter their beaks in greeting too, and the chicks learn very quickly to mimic this and join in when their parents alight at the nest with food.

Bowing to each other is another common greeting procedure, and this, along with beak clattering, features prominently in the elaborate dance performed by pairs of gannets. Black-backed gulls [39] have a similar ritual. Bridled terns [40] droop their wings and bob their heads in a form of bowing to each other in a pre-copulatory dance. The crested tern [41] offers a fish to its mate, while drooping its wings. The two birds then bob their heads to each other and rise into the air simultaneously in a straight-up spiraling flight.

In the courtship of the wandering albatross,[42] several males gather around a single female, bowing to her, their heads low to the ground, uttering groaning sounds which are evidently more attractive to the female albatross than they are to human ears, because she bows and groans back. After several bows the males half open their wings and sidle around her. Then they face her with wings wide open and curved forward, heads raised verti-cally, and utter loud cries, the whole procedure being re-

peated until pairing takes place. The couple build their nest together, and then they indulge in an altogether different dance
for their nuptial ceremony.

The Laysan albatross [43] has a mating dance in which the male
and female stand facing each other with their wings partially
open, while an absolute chorus or choir of their neighbors stands
around them crying out. The male and female raise their heads
with their necks arched, dip them to the ground, and, as they
raise them again, touch beaks. After simultaneously dipping
their beaks, so that they sweep under the right wings and then
under the left, the two birds lift their heads again and repeat the
whole performance. The repetition becomes more rapid, and the
noise of the other birds around them gets more and more intense
until it reaches a final climax.

Scientists cannot be sure, without careful experiments, to what
extent color or even the ritualistic movements are responsible
for the success of courtship in birds. It does seem that the latter
are the more important, because the drab female's response
movements are equally necessary for success; but in some birds,
especially birds of paradise, color, display, and song may be
equally important, for they all occur together, the display bringing to light exotic colors that would not be seen otherwise.
Sometimes sound alone must be a major factor in courtship. In
budgerigars [44] a soft warbling is associated with the male's pre-
copulatory behavior, and it is voiced while he is close to the
female. This apparently stimulates her ovary activity and her
egg laying; the longer the period of song, the more effective it is.

Some male birds feed their mates during courtship as well as
during incubation. This is more as a kind of symbolic ritual than
to provide nourishment, and it takes place just before copula-

tion. Some male birds actually feed their mates while copulating with them, and still others just after copulating.

Gifts of food are sometimes part of quite an elaborate ceremony. They are like gifts of candy between humans during courtship, or they may even be equivalent to the provision of a home and food in a human couple. Both strengthen the pair bond, and the female soon looks elsewhere if she is not provided for; or at least the relationship suffers.

Because courtship patterns are more extensive and have much wider variation among birds than any other class of animal, volumes have been written on the subject. For that reason, if for no other, limitations must be placed on the extent to which we deal with them in this chapter, but for those readers who want to pursue their study further, the Bibliography may hold a few suggestions.

3. The universal urge
in mammals

Polygamy and Monogamy

Although mammals have evolved the highest forms of sexual apparatus and the most advanced patterns of offspring care, they still enter into almost all kinds of family relationships—polygamy, promiscuity, and monogamy. Some of the lower orders, such as rats and mice, practice polygamy without any selection by the females. A female raises her young alone, the males passing on after copulation in carefree promiscuity. In bears both sexes appear to be promiscuous. The pronghorn buck [45] has a harem, which may number up to fifteen females, and so may other deer and antelope. The California sea lion [46] may have up to forty females in his harem, and the Pribilof seal [47] from forty-five to a hundred. Even the sedate koala [48] may control several females.

Monogamy is quite rare among mammals, although it does seem to exist among some of the canines and the American beaver.[49] Among some canines monogamy lasts for more than one season, but usually one season is the limit. The American beaver is one of the very few animals for which monogamy is permanent. But then he is so busy that this is understandable. Foxes and a few monkeys and apes are monogamous, but chimpanzees are most definitely polygamous.

When monogamy exists in higher animals, even in humans, sexual interest for the partner becomes increasingly less frequent and eventually may cease. Then the stimulus of novelty is required to produce a sexual reaction. At the other extreme from monogamy are animals that only copulate with each other, without forming any real partnerships. They live alone and meet for copulation, parting again for solitary hunting and living. There are even a few species, like the titi monkey,[50] that have stable partnerships, but will nevertheless copulate with members of other groups of the species when these are encountered.

OTHER MATING HABITS

Mammals do not seem to spend as much time in courtship as birds do. Many rely much more on mutual readiness to mate; and in some species preparation for copulation consists of nothing more than a male chasing a female, possibly making noises at the same time. Such chasing is quite useless if the females are not physiologically ready for copulation. There are species in which females chase males, but usually the female does not play an active part in addressing herself to a male at breeding time. She merely secretes an odor in her urine which is exciting to males.

The male and female Arctic fox[51] not only use odor attraction, but also indulge in unusual precopulation dancing in which they stand up on their hind legs face to face, looking rather like young cubs at play, but with more deliberation and grace. This is elaborate for mammals, which are ordinarily satisfied with a much simpler effort at persuasion. When the green acouchy[52] male is courting, for instance, he follows a female just wagging his tail furiously. If he feels bold, it will be vertical, but if he is nervous, it will be turned down. If she is interested, she shows this by raising her own tail vertically and arching her back downward.

In contrast to the energetic chasing and wrestling indulged in by some mammals, the elephant is a gentle kitten. When elephants are courting, they stroke each other with their trunks, intertwine these sensitive organs, and place the tips in each other's mouths, demonstrating a confidence and affection that is unmistakable in spite of the clumsiness of their bulk.

In great contrast to the elephant's sedateness is the swift dexterity of tiny bats, which try to copulate in flight and may even succeed, although there is no proof of this.

The guanaco [53] male is probably typical of many similar mammals; he merely circles a female and presses his neck against hers, nipping her and making grunting noises. He never moves far from her, because he has to guard her carefully against other males. If she has her way, she remains no male's exclusive property, and this keeps the male in constant competition, which in turn keeps him in good breeding form.

A few primates operate very much like humans in their approach to copulation, quite frequently the female taking the initiative, enticing the male to serve her when she is ready. To do this she presents her hind quarters to him with her tail raised, sometimes looking back at him and making sensuous sounds but at the same time simulating fear. When the male makes the approach, he may clasp her ankles with his feet, and grasp her buttocks with his hands, which makes it difficult for her to break away even if she decides to. The females of several other mammals take the initiative too. The female hedgehog lifts her tail and lays her spines flat in a gesture of invitation when she wants to mate.

In mammals with set breeding seasons, the male's season is longer than the female's. The latter arrive in the territory after the males have established themselves, and in some cases after they have made some preparation for breeding. This is very much like birds. In either kind of animal, fertilization is possible only when the female's ova are ready for it.

Both frequency and duration of copulation vary among species. Sometimes it occurs more than a hundred times in an hour for perhaps two hours with small rodents. With larger animals it may last from a few seconds to nearly half an hour several times a day. The blue whale [54] is very exceptional in that the pair leap out of the water into the air against each other, and in that brief moment copulation takes place. Mink [55] and sable [56] go to the other extreme. It is reported that they remain connected for up to eight hours.

Monotremes

The platypus mates in water after a short period of hibernation. The male takes the female's tail in his beak, and they swim in circles until eventually they copulate. Afterward the female retires to a breeding chamber which she has constructed in a burrow. It is separate from their living quarters, which the male occupies alone for the period when the young are raised. Little seems to be known of the actual courtship and mating of the echidna.[57] Its family habits are described later.

Male monotremes have poison spurs on their hind feet, which produce venom at mating time only. The echidna has been said to use these spurs hypodermically on the female to stimulate her, but we cannot be sure they are not more related to the greater need for protection or defense at this time.

Odor Marking in Placental Mammals

Most male mammals mark out the boundaries of their territory with odor from glands, urine, or feces, and this is repeated con-

10. The female platypus, *Ornithorhynchus anatinus*, constructs a special nursery chamber in her burrow, where she rears her young, and the male remains in the former living quarters.

stantly to keep the odor fresh. This is like birds' marking their territory with singing sites. Aggression toward trespassers is greatest near the center of the area, becoming reduced toward the boundaries. This is even true of animals that live in packs and mate within the pack or group, such as wolves, jackals, hyenas, lions, and coatis. Many males also use their odor to brand their females as their property or, as in the hamster,[58] to leave their signature all around a female's territory to keep other males away.

Black buck have large scent glands on the face, which they use both to mark out their territory and to brand their females. The male European rabbit[59] has quite a branding ceremony. He first makes a stiff-legged pretense at retreat in front of a female, keeping his tail flattened upward against his body, and then suddenly he rushes after her and squirts his urine over her. Porcupines, cavies, agoutis, and several others do the same thing.

Ring-tailed lemurs [60] have dark scent glands on their wrists, and there are others on the shoulders. They rub their tails on these glands to give themselves identity, so such odors must have a slightly individual character. Some female lemurs rub their urogenital regions on branches to mark them, just as males urinate on them for the same purpose. However, not all lemurs do this. Those that do are evidently marking their territory as well as identifying their sex. The potto [61] has special scent glands around the sexual organs, and scientists suspect that these too are used for marking and identification. The marking seems to

11. Black buck, *Antilope cervicapra*, have large scent glands on their cheeks, and the male identifies his females by rubbing these on her.

be stimulated by greeting or courtship and also by threat or un-
familiar things.

Just how these odors function is only now becoming clear.
They act, at least in many animals, as what are called "phero-
mones." These stimulate the opposite sex, and even induce estrus
and pseudopregnancy in females. This stimulation has been observed repeatedly in mice, as has the pheromone in the male
urine; and although it is not always wise to assume that what
is seen in the experimental laboratory will also operate in the
wild, it does seem that the pheromone is one thing that could
hardly vary because the same odor permits identification of
species, individuals, and sexes.

Male elephants have two glands near their ears, which at rut-
ting time secrete quite large quantities of strong-smelling secretion to interest females.

Color Signals in Primates

Many mammals, especially primates, signal their readiness to
mate by changing or heightening the color of certain areas of the
body, just as fish do. Female baboons, for instance, as well as
several other primate females, develop a prominent red rump
when they are ovulating. Sometimes this coloring is duplicated
elsewhere on the body, so that the information is passed to animals in front as well as to those behind.

The female gelada baboon [62] has a bright red skin patch around
her genitals, the central vulva being a deeper red, and the whole
area is surrounded by tiny white nipplelike projections. The
entire pattern is repeated on the chest, an area of naked red skin
being surrounded by tiny white papillae, and enclosing a pair of
vivid red nipples which have developed so close together that
they look just like a female genital organ. The color of all this
varies with the different stages of the estral cycle. The gelada

12. Mandrills, and some baboons, duplicate the colors of their genital regions on their faces. In the female the color shows her breeding condition.

spends more time sitting upright than walking on all fours, and so the message is conveyed, whatever the female's position; when she is walking, it is in the rear, and when she is sitting, to the front.

Primates that habitually walk on all fours have developed a different arrangement for carrying this message of sexual readiness. The female mandrill, for example, duplicates her genital coloring on her face, as does the male; the coloring of his bright red penis with blue patches on each side is repeated on his face in a red nose and blue-ribbed cheeks even more brightly than in the female. Although this facial coloring may give true sexual signals in females, since the brightness varies with ovulation, it is difficult to assume that this is true of males, which can mate at any time with full effect. So the coloring must have some social significance. In any case the male gives other cues to his sex, such as his greater size.

The Siamang gibbon [63] has a large rust-colored vocal sac, which is not ordinarily visible, but becomes so when inflated for calling. This may be a secondary sexual characteristic present in both sexes, but its color cannot be considered a signal of sexual readiness.

SOUNDS IN COURTSHIP

Many mammals use their vocal powers to augment their mating behavior. When female cats are ready for mating, the whole world knows about it, and the responses of the toms are no less enthusiastic. The copulation act is not exactly gentle either. The male grasps the female's neck with his jaws, and later when she has broken free, she strikes out at him with her paws. But one thing is certain, the female cat derives great satisfaction from copulation, and manifests this more than most other mammals.

Marine mammals vocalize a great deal. Porpoises have language for all purposes, and they would hardly neglect this valuable asset in courtship. Even seals call to locate each other at mating time. This call is loud enough to carry considerable distances, but it is quite different from the aggressive sounds made when a rival appears on the scene.

Most mammals find voices at mating time, especially for issuing warnings to rivals. Male deer of all kinds bellow their defiance to rivals; so do elephants, wild cattle, wild horses—and all are prepared to fight. Sounds used for actual courting are much less audible to humans at a distance. In fact, few of the smaller mammals can afford to advertise that they are off guard in their activities with such abandon as the wild cattle and elephant, which are more or less invulnerable. These softer sounds are certainly made, however, and they are frequently accompanied by tapping of the feet or clicking of the teeth.

DOMINATION

Not only may males dominate females, either singly or in

harems, but there are situations in which this domination extends to other members of the pack or herd, including all other males. This can have some interesting effects on the whole tribal relationship. For instance, one ram may boss or dominate many other rams and obtain a monopoly on breeding females. This influences the entire flock genetically, because most of the off-spring have the same sire.

The rams that are subdued by the boss ram become nervous and make little or no effort to breed; they merely witness his copulations up to six times a day. Furthermore, females that have copulated with one ram always cease to interest other rams, so everything appears to play into the desires of the boss ram. This may not apply to domestic sheep, of course, because of the controls exercised by man.

When a single male dominates several females or a whole group, the breeding season brings him many problems, one of which is that he may have very little time to eat. Seals and stags start their breeding seasons well fattened and in good health, but the constant guarding of their females from other invading males and continuous efforts to escape from the harem for weeks on end leaves them thin and exhausted.

4. Migration to breeding sites

Many animals travel considerable distances to and from their breeding grounds. In most the urge to do this is not always for breeding alone, but in fish it certainly is. Some return to their birth area to produce their offspring because this is necessary biologically. Salmon, sturgeon, some of the bass and herring families, and lampreys are but a few of the marine fish that must spawn in fresh water, where their ancestors bred, to ensure the survival of their young. Eels, on the other hand, which spend much of their adult lives in rivers and streams, must return to the sea to give their young the saline environment they need. These young hatch in currents that carry them to the-rivers where they live out most of their lives.

When birds migrate to breeding areas, they are in fact migrating for food as much as for breeding. Many birds, especially insect and seed eaters, would quickly starve and die if they stayed in one place, because of the seasonal nature of their food. So they follow the sun, demanding perpetual summer, even if this means traveling between the southern and northern hemispheres, as many do. Since most birds breed only once a year, and they always breed in the same general area, it does appear that they are migrating to their breeding grounds, just as fish do, but in most cases they are also migrating to fresh food supplies. Birds breed in the areas that are plentifully supplied with the kinds of food most suitable for the rapid growth of the young. These areas have become what one may call ancestral breeding sites.

The annual movement of animals such as buffalo, mule deer, antelope, zebras, bats, lemmings, and many others is always on the trail of food (or water) supplies. Their breeding locations are less tightly tied to a pattern than are those of fish and birds. Instead such movement responds for the most part to seasonal patterns.

Migration is not a simple response on the part of an animal to the knowledge that it should do this. There is no reasoning that tells a fish it should head for a certain part of the world to spawn. Nor is there any reasoning involved in a bird's migration to a place perhaps eleven thousand miles distant before its food supply dwindles. Animals migrate in response to a number of stimulating conditions, such as changes in body chemistry, length of day, temperature, position of the sun, each or all prompting them to do the right thing for survival. How these things work on the animals is only now becoming apparent.

Fish—The True Migrant Breeders

A rather awe-inspiring feature of the migration of fish over thousands of miles to their breeding sites is that they manage to negotiate with uncanny accuracy vast distances of featureless ocean from which they could have gathered nothing to be stored in their memory. Although experiments have given some answers to the questions this raises, not all the story is as yet obvious.

Perhaps the most important characteristic of fish that migrate for breeding purposes, either from salt to fresh or from fresh to salt water, is a difference in the chemical in the light receptors of their eyes. This "photochemical" changes its state when acted upon by light, and in the process it produces electrical currents, which pass to the brain and create the response known as "seeing." The wave pattern of these currents is different for different colors of light, and the peak response of this photochemical to light is different in freshwater fish from what it is in marine fish or in land animals. Because of its greater density, seawater passes different wavelengths of light than does fresh water. What then does this mean to the salmon and the eel?

Anadromus fish, which live in the sea but spawn in fresh water, and catadromus fish, which live in fresh water but spawn in the sea, have both freshwater and saltwater visual photochemicals, but one kind dominates the other. In the ocean fish drop to greater depths than in the rivers, and their eyes need to be more sensitive to the bluer light, which alone gets through to the deeper zones. However, although salmon spend most of their lives in the sea, their freshwater photochemical predominates, and in eels that of the marine fish predominates. This salmon characteristic is found in dozens of other northern ocean fish that ascend rivers to spawn; obviously, the dominant visual chemical in these fish is related to their birth environment and not necessarily to that in which they spend most of their lives.

This is part of the evidence demonstrating that salmon have changed their life habits quite drastically during their history. They were once freshwater fish, but have found it necessary to move to the greater food stores of the oceans in order to grow to maturity. The young are too delicate for this saline environment, and must still be given the opportunity to develop in the once native environment where the oxygen supply is greater.

But what of eels? The theory that springs to mind is that the sluggish-swimming adults are much safer in the rivers, where predators are less prevalent than in the sea. Eels too must return to their ancestral living environment to spawn. So far only one main eel spawning ground has been discovered in the Atlantic

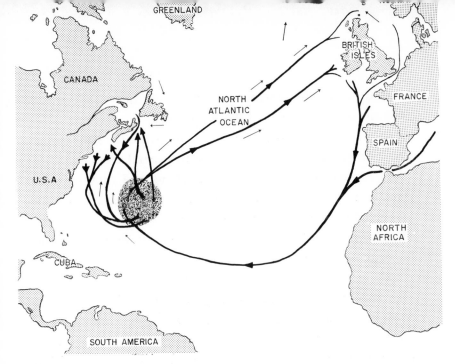

13. The spawning grounds of all Atlantic eels are in the Sargasso Sea area (shaded). From there the young elvers migrate to the rivers of Europe and America, always following prevailing currents, as shown by the small arrows. The larger arrows show the direction of their movements between their breeding area and the rivers in which they mature.

Ocean. This is in the Sargasso Sea, the area between the West Indies and the Azores, between latitudes 20-35 degrees north, at depths of 1,700 to 3,000 feet. At that level the eggs float motionless in water with a temperature between 61° and 63° F. (16° to 17° C.). The depth of the ocean at this point is actually 30,000 feet.

Both the American and European eels return to the same approximate area to spawn and die, and their young seem to return to the parents' native rivers. The European species evidently breed on the fringes of the Gulf Stream, and the young are probably carried the 6,000 miles to Europe in the stream. The evidence for this, however, is not conclusive. They take three years to cover this distance, whereas the American eels take only one year to make their shorter journey.

The eels on the two continents are not from the same species, even though they all spawn in the same general area. Those that reach Europe have 114 or 115 vertebrae, and those reaching American rivers have 107. This may prove also that the two species breed in different parts of the Sargasso, because the number of vertebrae in fish differs when the temperature in which they incubate is different.

Eels in Australian and New Zealand waters have the same migratory habits, and the breeding ground for these appears to be in the Tonga region. Only two other eel spawning grounds are known. One is in the western North Pacific, and the eels that are born there migrate to and from China and Japan. The other is in the Indian Ocean.

Before leaving the rivers for the sea to breed, the female eels change from a dark color to silver, and an even more miraculous change takes place in their eyes. These grow immensely, and the

14. The breeding area for Pacific eels that migrate to and from Australia and New Zealand is near the Tonga.

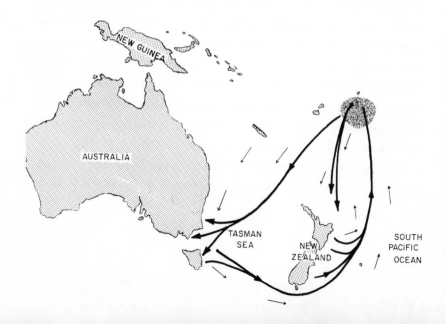

eels' ability to use extremely sensitive light receptors to see in the darker depths of the ocean improves. The baby eels born at these depths are also silver and have very large, sensitive eyes. Their eyes gradually become reduced, however, as they approach the rivers where they eventually settle as the dark-colored lethargic creatures we know as mature eels.

When fish return to their birthplaces to spawn, they are not just migrating; they are "homing" as well. Experiments have shown that at least some species may use the position of the sun to orient themselves on their migratory journey. Tuna especially are said to travel great distances in this way, across both the Pacific and Atlantic oceans, but there are many questions to be answered before we can assume this as infallible. We only know that migratory fish swim more randomly in cloudy weather.

There is a theory that as spawning time approaches, the bodies of anadromous fish demand more oxygen, which is only available to them in river water. The higher upstream they go, the greater the oxygen content of the water is. This theory is not proven, but as already mentioned, more oxygen may be necessary for the incubating eggs and the fry.

Once migrating salmon are in the main river which they first traveled down to the sea as young fish, they may find their way to the original tributaries and small streams solely by smell. Experimental work supports this theory. The birthplace imprints a characteristic smell in the salmons' memory. This may be a little difficult to accept; but the experiments are reasonably convincing, and we know that their noses are so sensitive they can detect a few molecules of a substance. We know too that in many animals recognition is by the sense of smell.

The soil chemicals and the plant life through which the birth stream passes may give a characteristic smell to the water, which can be recognized by the fish's sensitive nose. Plugging the noses of these fish when they swim upstream certainly makes them choose tributaries in an entirely random fashion. This theory may not be as likely to apply to fish that merely return to river estuaries or tidal regions, although even then it remains a definite possibility.

The eel also rather confounds this idea, because it is born far out and deep in the ocean, migrates to rivers and streams, and then returns to its birthplace in the ocean to spawn and die, covering maybe thousands of miles of featureless water. No memory of smells could possibly provide a trail for the eel to follow to its final destiny. And there are other fish that live in fresh water and spawn in the ocean, such as the New Zealand whitebait and a smelt in the Falkland Islands area.

The argument that an odor trail is used in migration is supported by newts that have had their eyes removed and been transported considerable distances from their home territory. These animals can still find their way home, even though it may take them a long time, and they can accomplish this only by their sense of smell.

There are numerous species of salmon in the Pacific and the Atlantic oceans, all having interesting differences in their breeding cycles. The five species of Pacific salmon on the American coast, and those that migrate in the opposite direction to Asia, all mature at sea, taking from two to eight years. They spawn once, after fighting their way upriver for distances of up to 2,000 miles, without feeding for weeks, and overcoming seemingly insurmountable obstacles such as rapids and waterfalls, which they leap up, sometimes to heights of 10 feet. After spawning, they die from starvation, exhaustion, and loss of calcium. In the Yukon River, which is no placid stream, the salmon

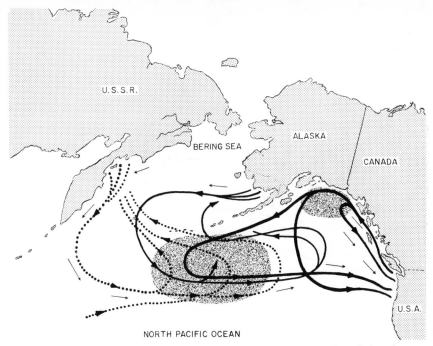

15. Several salmon species live in the Pacific Ocean, and each has its own breeding rivers. Some go to Alaska and the Pacific northwest of America; others go to the Siberian coast.

cover 1,500 miles in thirty days; the exhaustion from this must be phenomenal, because 50 miles a day is probably twice as many miles as most other salmon cover.

The hardiest of the Atlantic species do not all die after spawning. Some return to the sea, and they may find their rivers to spawn again up to three or even four times. There are some lake and landlocked varieties which, although they cannot escape to the ocean, still simulate the spawning cycles of their free relations. Naturally, with their more restricted food resources, these do not reach the same size at maturity as the ocean-living species.

Migrating salmon are able to endure the long period of starvation because their flesh contains 17 percent oil, and this oil is a fat storehouse, which can be drawn on during the journey. If, by the time spawning is completed, the oil is all used up, body

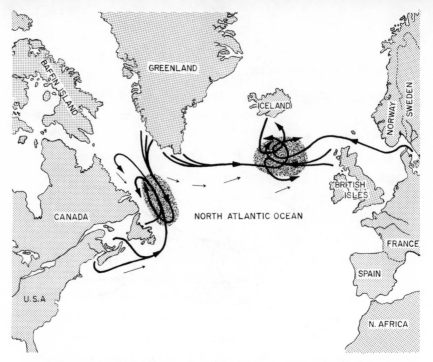

16. Atlantic salmon do not migrate as far as those in the Pacific, and there are two main areas where they may concentrate before moving toward their rivers and streams in Europe, North America, Iceland, and Greenland.

wastage exceeds 40 percent, and the bones have become soft through loss of calcium, the fish die. Possibly, those Atlantic salmon that spawn more than once are able to do so because their food reserves are not entirely used up. They negotiate much shorter rivers than the Pacific species, and they can get back to the ocean, where calcium carbonate hardens their bones again, before these have degenerated too far.

Like salmon, lampreys spawn high up the rivers, and some species then descend to the sea to mature. This tells us that, at some distant point in their history, these fish also changed from a life in fresh water to one in salt water. This appears to have been a general move on the part of fish between 300 and 400 million years ago, but some of them have continued to return to their birth environment to spawn.

Whatever the factors that stimulate fish to migrate to their spawning grounds, quite obviously sexual maturity must be one of them. This, in turn, is linked to the increase in the length of the days and an accompanying temperature rise as well as to increased activity of the pituitary and other glands. (See Chapter 5.)

The three-spined sticklebacks [2, 64] move from salt to fresh water to spawn in spring, a movement that can be stimulated at any other time by treating the thyroid gland with glandular extracts. The physiological change that takes place at this time gives the stickleback a preference for fresh water, and in the autumn the very opposite takes place. Experiments in which the day length was artifically shortened have shown that when the days are short, the stickleback does not acquire a preference for fresh water, even when the season is right. Lengthening the day, however, can induce the preference quite prematurely.

Juvenile Pacific salmon swimming down to the sea in spring show the same preference mechanism. Even those species that are landlocked and cannot escape to the sea show a change in preference from fresh to salt water at the end of the migration season. All the evidence indicates that this is due to the effect of changing day length on the body glands.

BIRDS

Glandular changes also take place in birds when the day length reaches critical phases, but these are not always concerned with breeding or changes in the gonads; sometimes they are concerned only with the urge to migrate. Not all birds breed during the first year of life, so their first migration lacks that impetus. This is confirmed by experiments in which birds have been castrated. They still migrate with their fellows.

In no branch of the animal kingdom are there so many migrants as there are among birds (about one third of all birds). Nor do any other animals travel so far in a year. The Arctic tern [65] may cover up to 22,000 miles or more in a single year flying between the summer and winter territories. One that was banded on the Arctic coast of Russia was caught off Australia, 14,000 miles away.

Pacific golden plovers [66] breed in Siberia and Alaska, after migrating there from as far afield as Argentina, New Zealand, and Hawaii. The white-rumped sandpiper [67] flies from Canada to Tierra del Fuego. Some barn swallows breed in Alaska and migrate 9,000 miles each way to and from Patagonia. The bristle-thighed curlew [68] commutes 6,000 miles each way between Alaska and Tahiti.

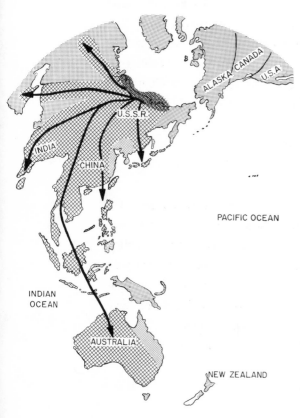

17. The curlew sandpiper, *Eurenectes ferruginea*, breeds in the Arctic areas of northern Russia, then travels in many directions when the northern summer draws to a close. The cross-hatched areas show where it is found, and the arrows show its migratory routes.

Birds mate and rear their young between arrival and departure in a particular area, so their migration has long been linked in people's minds with breeding; but as was mentioned at the beginning of the chapter, it is highly unlikely that the two are actually linked in all species. The curlew sandpiper is one species that may be like the salmon, however. These birds nest and rear their young in the far north of Europe and then, as winter approaches, scatter to India, Africa, Asia, and even Australia. Although they arrive in these places for their summer season, they do not breed there. They breed only in the one northern area.

The reason for migration, then, is not always just to find a suitable breeding territory, but also to find food. Few tropical birds travel far; they do not run out of food. But in areas well away from the tropics, where there is a winter period when neither insects, fruit, nor flower nectar are available, movement to another area, where either it is summer or there is a perennial food supply, becomes essential for survival. Even some predatory birds, such as peregrine falcons, which live in Arctic regions, must migrate southward as the northern summer ends and the ice and snow close in over their hibernating prey. This is easy to understand, and so is the migration of many hummingbirds from one flowering season to another, although why it should be necessary to fly 2,000 miles for this, as the tiny ruby-throated hummingbird [69] does, is not quite so obvious.

Some seabirds make their headquarters for breeding in a certain place, and then for the rest of the year range the oceans freely, following routes that cover good feeding grounds. They finish back at the breeding site, using prevailing winds to take them on their circular route.

To the extent that migration in birds is not exclusively for breeding purposes, as it is with fish, we should perhaps not be so concerned with it here; but we are concerned for another reason.

How do the young learn to migrate, and how do they know when and where to go? Is this part of parental training, or is it some built-in mechanism? If it is parental training, then we need to examine it in the context of our subject.

We do know that many young birds travel in their first season with experienced adults. However, there are also some birds, such as the gannets and the petrels, that follow later without adult guides. Others do not travel in flocks at all. They take off in several directions, individually, and only gather together again in the next breeding season. These birds could well memorize the route they took when they scattered.

Experimental work has revealed that some birds, like the golden plover and the pigeon, are able to use the sun to orient themselves during the day, even making corrections for its movement across the sky, and to use the stars at night. But they cannot do this if they have not been allowed to see the horizon during their growth. Ducks can use features of the landscape, the sun, the stars, and may even be able to sense the directions of the prevailing winds to establish the direction they are heading.

Birds that travel over broad expanses of ocean could not possibly find their way visually except by using celestial bodies. Nor can they drop down and wait for clear days and nights if clouds suddenly obscure the sun and stars for some time, so obviously this form of navigation is not the whole story. Since migratory routes are upset by radio waves, it has been suggested that birds respond to the magnetic pole of the earth, using this response as an orientation cue, and that this response is disrupted by the radio waves. So far this idea has not been widely accepted.

There is, however, a mechanism that sometimes appears to break down spontaneously, because each year some members of a species do not respond to the urge that stirs their fellows.

These are usually picked off by predators, or they succumb to climatic conditions. They seldom survive in the severe regions. Birds are also said to orient themselves by the use of polarized light, and experiments have confirmed this in some species.

Ordinarily birds must feed constantly. How then can some of them travel for days, even weeks, over hostile land and ocean with neither food nor water? Some are known to stop and feed, and it is suspected that many more may do so, because frequently those that cross oceans keep a coastline in sight or hop from island to island. This may be for both feeding and route identification, as in birds that cross continents by following river valleys. Certainly some appear to do all they can to avoid crossing large bodies of water devoid of identifying features.

The pectoral sandpiper [70] breeds in the Arctic and migrates there from South America without ever losing sight of land. The bobolink [71] breeds in Canada and also commutes to South America by island hopping across the Caribbean. Storks, which breed in Europe and Asia and migrate to South Africa, never go out of sight of land. They either cross the Mediterranean at Gibraltar or go around Asia Minor, a journey of 8,000 miles.

Those birds that follow very long routes on which they cannot feed, store fat to supply the energy needed for the journey. Before leaving for the winter territory, they feed avidly and build up vast fat reserves to be drawn on during flight. Fat is one of the lightest body tissues, but ordinarily it contains a lot of water, which is relatively heavy. In migratory birds, however, this fat is concentrated to exclude much of the usual water content, and it is not located in the muscle and other body tissues, as it is in mammals. Instead, it is stored in spaces between tissues and organs, whence it can be drawn to supply energy fuel without affecting the weight or bulk of the muscles used for flying.

Only in this way can birds make these long, arduous flights and still arrive at their destinations looking fresh, energetic, and well fed. Then, too, besides being the food that provides the most energy, fat, when it is burned by the body as fuel, manufactures water, which is just as important. In this way the birds manage without replenishing their water also.

The albatross and the shearwaters typify the birds that roam freely over the oceans throughout the year and then return to specific sites only to breed. The route of the muttonbird, a short-tailed shearwater, is shown in Figure 18, and a close

18. The muttonbird, *Puffinus tenuirostris,* breeds around the Bass Strait area of Australia, and then takes off on a mighty ocean ramble that circles the Pacific ocean before it returns to breed. The little stint (dotted-line route) flies from Australia and Malaysia in March and April to Arctic Russia and Alaska, where it breeds before returning in September and October—a round trip of 20,000 miles. The fine arrows show prevailing winds.

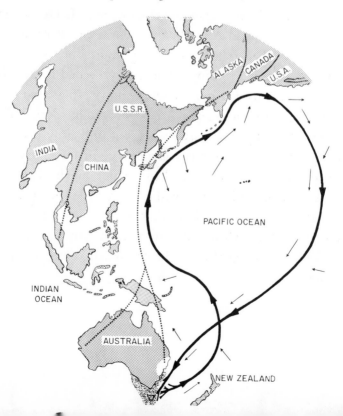

study of the prevailing winds will show that these birds evidently ride the winds all the way around from Alaska to South Australia. Do they, then, merely use these winds to carry them along, or have their migration routes become set because they follow wind paths anyway? Perhaps it does not matter, so long as they do reach their breeding grounds.

Birds such as the greater shearwater do not migrate or return in organized flocks. But when ready, they leave Tristan da Cunha and the Falkland Islands in the direction of South Africa, turn north and cross back diagonally over the Atlantic to the North American continent, thence down its eastern coast to their starting point again, riding prevailing winds for much of the route, and miraculously gathering together after being widely scattered for months. As with the petrels and gannets, parents leave before the young are able to, so the latter have no visual contact with experienced migrants.

The royal albatross [72] may not touch land for nine years after leaving the nest, so prevailing winds can hardly bring this bird back to its birthplace. Even gannets do not always breed until their fifth year. In their case, however, younger birds do return and, in many cases, stake out and defend a small site, but without breeding.

Maybe experience is not required to ride a route on prevailing winds, but experience must be important to many species, because parental or flock guidance is practiced. Some wild geese travel as a family, juveniles with their parents, and many family groups making up a flock. Some travel at night and keep in touch by constant calling, an arrangement that enables them to feed during the day.

The release of the sex hormones and the gonad development of migrating birds heading for their breeding grounds coincides with a change in day length. Just as these changes play a

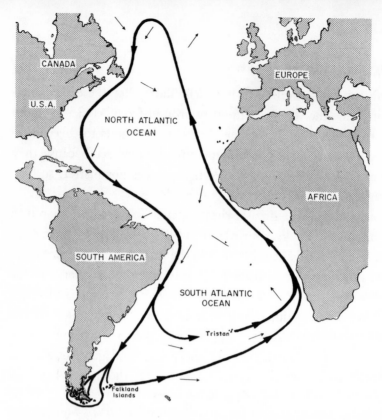

19. The greater shearwater, *Puffinus gravis*, breeds on the Falkland Islands and Tristan da Cunha, and the sooty shearwater breeds around the tip of South America. Both species roam north and south over the Atlantic, as the muttonbird does over the Pacific. The fine arrows show the prevailing winds.

part in prompting them to leave for their breeding territory, changing day length and subsiding glandular activity at the end of the breeding season have the opposite effect. These changes stimulate them to leave for the areas they occupy during the rest of the year.

Penguins also migrate for breeding purposes, either by swimming or walking, and at least some of these present us with another interesting behavior pattern. Adélie penguins [73] migrate in a more or less straight line across hundreds of miles of featureless Antarctic ice to their breeding sites. How? And why? The "how" can perhaps be explained by the fact that their movement is disrupted on cloudy days, indicating a dependence on the sun for orientation and an inbuilt biological clock enabling each bird to compensate for the movement of the sun in the sky. The "why" is not as easy to explain.

Emperor [74] and king [75] penguins breed during winter, when it is dark, everything is frozen to a temperature of 70° F. below zero, and there is no access to food through the ice. These birds, then, must be able to go without eating for long periods, since they incubate their eggs in the extreme cold. So perhaps the Adélie penguin is not searching for a warmer place to breed when it waddles for miles over the ice, but rather for the nesting materials it demands. (See Chapter 5.)

MAMMALS

Some bats make long migrations to breeding sites, but they are probably not migrating for any reason other than to obtain new food supplies. Most bats that live in nontropical regions and inhabit caves or holes hibernate for the winter; but those that habitually hang in trees cannot do this, and so they migrate to warmer climates and other feeding areas. The red bat [76] and the silver-haired bat [77] fly for distances up to 700 miles, even across sea, in their migrations. They seem to use prevailing winds, following the coastline at night, some distance from land, and often landing on ships.

Australian and African flying foxes and tree-roosting bats migrate solely to new sources of fruit and flowers, and can probably bear their young wherever they happen to be at the appointed time. But many bats migrating in Europe return to precisely the place they left the previous season. One tiny pipistrelle is known to migrate well over 700 miles, not far by bird standards, but far beyond visual limits.

The migration of bison, caribou, etc., is evidently for food that remains available in less extreme climatic conditions. Certainly, their return to summer grounds is in response to the renewed availability of vegetation. Whales migrate between the tropics and the subpolar latitudes, giving birth in the warmer zones. It seems possible that whales migrate in a way similar to what shear-waters do, taking in the widest feeding areas. There are female fur seals, however, that migrate alone for 3,000 miles from Alaska to southern California, in a true breeding migration. We still have much to learn about the migrating mammals.

It seems that marine mammals use their echo-sounding skill, together with the natural sounds of an area, to test the topography of the ocean floor and the coastline, and to locate their whereabouts and direction when migrating from one place to another.

None of this entirely answers the question of whether or not migration to breeding sites is part of parental leadership or training. Obviously, in many instances it is not, but when both experienced and young adults travel together over the migration route, it does seem to be. Again we encounter a phenomenon with too many variations in it to permit us to generalize.

Part II

The Responsibilities of Parenthood

5. Preparation for family life

In the animal world the building of a nest or burrow, or even a house, is an important and significant step in the preparation for family life, as permanently imprinted in an animal's habits as its actual courtship.

There are some excellent house builders in the animal world. From the simplest hollows in the ground to the most elaborate apartment blocks, the results are all achieved with no more tools than teeth, beaks, claws, and tails. Complicated tunnel systems evade flooding or provide escape routes from raiders, have camouflaged entrances and exits, include nurseries, living quarters, and sleeping chambers. Although in many species much of this is for general living, it is perhaps even more related to the rearing of young and the protection of the family. Some buildings, as those of mound-building birds, are solely for the incubation of eggs, but always there must be an element of accuracy in construction that took many thousands of years to learn.

With the provision of living and breeding quarters, we see much more actual pairing, although not necessarily more monogamy. In fact, one sees repeatedly that monogamy is a rarity in the animal world in relation to the number of species there are. Perhaps the reason monogamy is rare is that it may depend very greatly on competition between females, and there are relatively few species in which this has evolved. It is well established in humans, although this is skillfully camouflaged by the females themselves.

NURSERIES, NESTS, AND INCUBATION HABITS IN FISH

Even the lowly lampreys include species that fashion depressions in the riverbed to receive their eggs. This is the vertebrate world's first kind of nest. Many other fish do the same, but there are some that actually build enclosed nests from plant material. The stickleback does this, as was described in Chapter 1.

The paradise fish,[3] the Siamese fighting fish,[4] and the gouramis, *Trichogaster* spp., make surface nests of bubbles, which they create by taking in air and expelling it encased in a water-resistant secretion. The eggs are laid into this, and the young remain in it at first; then they hover under it until they take off for safer places. Another gourami, *Osphronemus goramy*, builds a nest of mud, plants, and roots, which takes about a week. Then the female lays her eggs deep inside, after which both male and female fan and aerate the opening until the young hatch about ten days later.

Copeina arnoldi, a characin found in tropical fresh waters, manages to stick her eggs to a rock above the waterline. Here they are safe from other fish, and the male stays below them and splashes them continuously to keep them wet until they hatch and fall into the water. In another species both male and female repeatedly leap together from the water and stick their fertilized eggs to leaves or twigs overhanging the water. Then again the male splashes them for about three days until they hatch.

AMPHIBIAN NESTING AND INCUBATION HABITS

A few amphibians construct nests, some even on land; but.

many more lay their eggs in sacs or in masses, or wind them around plants. Some even carry their eggs around in special receptacles or structures on their own bodies. Amphibians have probably adapted to the widest range of hatching habits of any group of animals.

Two tree frogs, *Hyla rosenbergi* and *Hyla faber*, build basins of mud near the edges of pools or even in pools. One salamander, *Desmognathus fuscus*, lays her eggs in excavations in soft earth beneath stones or logs close to water, and the young are able to remain in these for fifteen or sixteen days. Another salamander, *Plethodon cinereus*, lays her eggs in crannies in logs, where the young hatch and leave as miniature adults; and a related species, *Plethodon glutinosus*, lays its eggs deep underground in the walls of caves.

One unique salamander, *Ambystoma opacum*, lays her eggs on land and just lies curled around them until they hatch in the rains and the young make their way to the water. The female *Amphiuma* also curls around her eggs. There is an African toad, *Hemisus*, that lays its eggs in burrows it digs, and the young find their way to water after hatching. The majority of the amphibians, however, do not make true nests.

Many newts attach their eggs singly to waterweeds. One American species, *Triturus viridescens*, wraps a leaf around each egg, but if there is no vegetation available, she will just attach them to stones. The male midwife toad [78] draws ropes of eggs from the female's cloaca and twists them around his thighs, often drawing eggs from several females. When the young are about to hatch, he goes to the water, and they escape into it. Some South American tree frogs carry their eggs in masses on their backs, and one, *Gastrotheca*, uses his legs to direct the female's eggs into her pouch. The tadpoles grow in the pouch and receive their nourishment through special outgrowths that interchange with her circulation. When the young are ready to escape, the female opens the pouch with her hind legs and frees them into the water.

When the male Surinam toad [79] mounts a female, his tight grip forces a tubelike string of eggs from her, which the pressure of his body directs over her back. A hormone is released, which stimulates a layer of skin to grow over them, forming a little cell for each egg, and each of these cells has a lid. Within their cells the young absorb, probably through their tails, a protein secretion given off by the female, and at the end of the incubation period fully formed miniature toads break through from their individual cells.

Instead of carrying their eggs themselves, the females of some tree frogs stick them to the backs of the males, and when the young are developed, they swim away as he enters the water. Others have only slightly different arrangements; the tadpoles of two species, *Phyllobates* and *Dendrobates*, are carried on the backs of the males to streams where they complete their metamorphosis instead of on the parental backs.

One South American frog, *Rhinoderma darwinii*, is perhaps the most unusual of all. The male carries the eggs in his vocal pouch, where they hatch and the larvae metamorphose into adult forms before leaving. The female lays twenty to thirty eggs, as several males gather around to spawn. These males fertilize and guard the eggs for two weeks, by which time the tadpoles are beginning to show slight movement within the eggs. Then each male tries to take as many eggs as possible into his mouth and thence into a throat pouch, which extends almost from his chin to his thighs. There the eggs connect with his body tissue to derive nourishment. Finally, miniature fully formed frogs emerge from his mouth.

There are a few frogs that, like the paradise and similar fish, make foam nests in water, releasing eggs and sperm into these. The females stay on guard until the young hatch.

REPTILES

Many egg-laying reptiles bury their eggs in holes in the ground. Those lizards that lay eggs probably hide them most often beneath or among stones or debris or in a hole in soft soil, where they develop for six to twelve weeks before hatching. The sand goanna [80] does not make any kind of nest, but it does dig a burrow several feet long, which perhaps amounts to the same thing. The eggs are laid at the end of the burrow, and it is then filled in. The thorny devil [81] also lays its eggs in a burrow, and the young eventually scratch their way out one night and greet the world at dawn.

Iguanas burrow into termite nests, and there the young have a ready food supply when they hatch out. The termite mounds may also supply the right amount of heat to incubate the eggs. Many reptiles use a pile of rotting vegetation as a nest, getting both moisture and heat from the humus.

Some snakes use rotting vegetation, and pythons make a nest of their body coils. Only one snake is known to build a true nest for its eggs, and that is the king cobra. [82] This seems strange in an animal that grows to a length of eighteen feet, is highly venomous, and has enough strength to reach and dominate almost any protected place.

Sea turtles lay their eggs in holes that they dig in the sand of a shoreline. A few freshwater turtles and tortoises possess large bladders connected to their intestines. They keep these filled with water, which supplies them with oxygen, and they empty water from these bladders onto digging sites in riverbanks, which makes the earth soft so that they can excavate nests.

Some crocodiles just bury their eggs in sand, soil, or debris;

20. The wagtail's neatly cemented nest is securely held and safe from most predators. It is just large enough for the setting bird to seal the top and completely cover the eggs within it.

but the estuarine crocodile [83] builds a nest five or six feet in diameter and two feet high with grass, debris, and mud, which the female keeps moist with urine.

BIRDS

First prize for nest building must naturally go to birds, which have designed, perfected, and learned to fashion the most intricate nurseries one can imagine. There is a very good reason for this. Birds must raise their young in some very precarious places, so if possible they must satisfy two requirements. First, the nest should be located where it is reasonably safe from predators, and second, the young should not be able to fall out of it.

Just as every other class of animal has species that burrow into the ground, some birds, including the puffins, petrels, and a few penguins, have adopted this kind of security from predators. The parrots and woodpeckers use holes leading into hollow trees.

Not only are the birds the great nest architects of the animal world, having perfected the tiny one-inch nests of humming-birds as well as the great incubating mounds of the Australian megapodes and scrub fowl, which sometimes rise twenty feet into the air with a diameter of up to sixty feet, they also work the hardest to achieve what they do. The bottle-shaped nest of the fairy martin is made of as many as 1,300 individual pellets of mud, each one requiring a trip to dirt and water, careful mixing, and laborious transport back to the nest site.

21. The pied cormorant's nest, which the bird may put almost any-where, is not nearly so protected from poaching as the nests of many other birds, but this is a large bird and can protect its eggs.

22. The nest of the fairy martin, *Hylochelidon ariel*, requires up to 1,300 mud pellets. The bird makes each one by mixing dirt and water, and then carries it to the site.

23. This Cape Barren goose is satisfied with using the grass at hand to make a rough nest in a shaded spot.

For some birds one of the important features of nest building is camouflage; so the sites chosen for nests are as variable as the habits of the birds themselves, and certainly the nests are varied to blend with their surroundings. Grebes build floating nests of waterweeds and rushes, and birds such as terns and dotterels merely scrape out a depression in a sandy beach or stony ground, just about as invisible as any leaf nest in a tree.

Camouflage is not always sought, especially by powerful birds like the wedge-tailed eagle [84] and the European stork,[85] both of which build great stick platforms. It would probably be difficult to disguise anything so large anyway. The tiny plaited purselike nests of some flycatchers cannot be seen unless really searched for, their size alone being almost enough to camouflage them from predatory eyes.

Kingfishers burrow six or seven feet into riverbanks for protection. Weaverbirds build hanging nests which a heavier creature could not approach without risk of a fall; swallows build strong nests with clay. Most swifts use saliva and sticks, feathers, and other materials to make their cement nests. The palm swift sticks its eggs to a leaf instead of building a nest, and there are birds that build where hornets will protect them.

Tree hollows are favorite sites, and in areas where these are scarce there may be bitter competition for them. In a mallee district, where large trees are rare, a hollow tree may be occupied successively by pairs of galahs, cockatoos, parrots, nightjars, and kestrels. Tree-nesting ducks have problems finding these scarce holes too. As soon as one duck brings out a clutch of young, another female hops in and brings up her family, building on top of the previous nest. Often the traffic in these trees is so thick that many different ducks' eggs get piled up layers deep.

The lyrebird has a most varied range of nesting sites. The female builds the nest, and it may be on the ground, on a rock

24. Parrots, cockatoos, and galahs all like to nest in hollow trees or holes, and the scarcity of holes in some areas makes competition for them very intense.

ledge, on a shelf in a cave, or even high up in trees. It is often thirty inches in diameter and fifteen inches deep, with a domed inside chamber ten or eleven inches in diameter and an entrance at the side just large enough to admit the female. She builds it with twigs, lines it with fern roots and other vegetation, and plucks feathers from her thighs as a final lining for warmth.

The hornbill female really cloisters herself. She nests in the hollow of a tree, lays her eggs, and while she is inside seals the entrance with excrement, mud, and wood brought to her by the male. All that is left is a small hole through which she is fed by the male. While she is sealed inside, she molts, and the feathers line the nest. Most Australian parrots nest in hollow trees or stumps, and three species nest in termite mounds, only a few species being ground nesters.

What architectural contrasts there are between the female emu's simple ground nest of grass, leaves, and bark, and the nest of a pair of eagles found in Florida measuring nine and a half feet in diameter and twenty feet deep. Still another contrast is presented by the nest of a tailorbird, *Cisticola exilis*, which is made by drawing leaves together and stitching them with fibers or by weaving fiber threads around grass stems.

Brush turkeys in Australia, *Alectura lathami*, build mounds like crocodiles, and some as large as 150 feet in circumference have been recorded. The temperature within reaches 90° to 96° F. Several hens may lay their eggs in one mound, and the male opens the mound and tests the temperature with his beak while the eggs are incubating. If the temperature is too low, he adds more leaves to the mound; if it is too high, he opens the mound up to some extent. When the young hatch, they must scramble through the mound before they can run off independently into the forest.

The male mallee fowl [86] builds and tends his five-foot mound and makes the final decision as to when it is ready for the female to lay her eggs. It must be 92° F. within. She lays from six to thirty eggs, and incubation is usually complete in fifty to eighty days, depending on conditions. When these chicks hatch out, they also head straight for the bush or scrub.

Not only do male mound birds protect their nests, air them, cool them by digging holes in them, and increase their heat by covering them with more debris, but the males of New Guinea brush turkeys, *Talegallus* spp., cooperate in building a mound up to sixty feet in diameter and ten feet high. When this is completed, each excavates his own incubating chamber and brings his own female to fill it with eggs. On Savo Island in the Solomon group, there is a short-tailed megapode that nests close to the shore and merely kicks quantities of earth over its eggs. When the young birds scramble out about forty days later, they fly immediately, if hawks do not get them.

Gentoo penguins [87] also make mound nests with vegetation, which they erect on a foundation of stones. Where vegetation is not available, feathers are used on the stones. Adélie penguins,[73] which are found farther south than the gentoo, also use stones, and build a rough wall around the nest. When stones are in short supply, a great deal of time is wasted in stealing them from other nests and egg stealing even occurs if a clutch size seems inadequate to the parents.

Both parents share in the incubation and feeding of the young, one sitting continuously for the first half of the incubation period, and the other for the second half. The jackass penguin,[88] which breeds in caves and uses sticks for a nest, has the same shortage problem, and at such times also becomes a thief, stealing from its neighbors. The theft of nesting materials may be more common than one would think, because certainly this has been observed in several bird species. In fact, some have been seen to dismantle other birds' nests for the materials they contain. Others just take over the nests other birds have built so laboriously, a kind of hoodlumism that is fortunately rare.

Several African birds, especially those that weave hanging nests, build these in huge colonies covering entire trees. A parrot in Argentina builds a huge community apartment block, in which each chamber is separate. The whole structure may eventually weigh more than five hundred pounds and fall to the ground because of its own weight. One of the weaverbirds also builds a large apartment house, twelve or fifteen feet in diameter, in a tree, and this may hold as many as a hundred pairs of birds.

Some South American cuckoos also build these communal nests and breed as a group. Just why so many species of birds do this is not clear, however, because there is one distinct disadvantage. The food supply in the immediate vicinity of such a bird village is depleted quickly with so many small mouths to feed after the young hatch.

It is interesting to see the caution (we would call it wisdom in humans) that seems to be instinctive in birds when they choose their nesting sites. Many water birds and birds that nest near water hold back on their building until the water has reached the highest point that it will rise to during the season. This not only avoids nest flooding, but ensures that most of the water life with which the birds will feed their young is well developed. If rivers flood or overflow their banks, ducks, ibis, herons, egrets, pelicans, cormorants, water hens, and a host of others assemble and wait to begin nesting as soon as the plant and animal life is adequate.

INCUBATION BY BIRDS

The incubation of eggs does not quite fall under the heading of this chapter; nor does it fit into the subject of "Giving Birth" or "Being Parents," as the next two chapters are entitled. So, since it is a period preceding actual parenthood, this is where it must be included.

Most setting birds develop a "brood patch." This is an area on the breast devoid of feathers, so that the bare skin can come into contact with the eggs and transfer heat to them; extra blood vessels form in the area, conveying additional heated blood there. The gannet does not develop a brood patch, but instead places its large webbed feet over the egg. The blood vessels in these keep the egg well heated, while the body resting on the feet insulates them from heat loss in any other direction.

So long as a parent bird does not desert its eggs and turns them regularly to ensure that they are evenly heated on all sides, the embryos develop normally no matter what the weather may be. Some birds turn their eggs every few minutes, but much depends on the air temperature. When this is favorable, the eggs may be turned about once an hour.

Penguins have real incubation problems, because of the cold regions they inhabit. The male emperor penguin [74] carries the single egg on his feet, covering it with a large fold of skin and fat, which hangs from his abdomen. He cannot feed during the incubation period and may lose as much as twenty-five pounds. The king penguin [75] does the same except that both male and female share the incubation task, rolling the egg off the feet of one onto the feet of the other. As if this were not demanding enough for parenthood, the female must go hungry for two months and the male for four months. She goes first on the long journey to the sea to feed, while he stays with the egg. Sometimes the Adélie penguin lays two eggs, and when this happens, incubation is even more demanding. The penguin must put one egg on its feet and the other on the ground or nest, covering the latter with its body.

Penguins are not the only birds that go hungry during incubation; the male emu goes without food for sixty days at this time. Some other birds seem to arrange their affairs a little more conveniently. While the female budgerigar is setting on her eggs, the male feeds her regularly by regurgitation.

Keeping eggs warm is a constant problem for some birds, but there are places and times when some have the opposite problem: keeping them cool. Some birds stand in water to wet and cool themselves and then sit on the eggs in torrid heat, but although this cools the eggs too, it is unlikely that they do it by any reasoning. They are probably trying to keep their own temperature down.

MAMMALS

Monotremes, marsupials, and placental mammals use nurseries in burrows most often, instead of using nests in the way that birds do. The platypus lines her nursery with wet leaves and lays her eggs on these. The tunnel to the nursery may be any length up to sixty feet, and when she retires to incubate her eggs, she seals this behind her in several places.

The house-building rat does emulate birds to some extent, however. It uses sticks and debris to build a complex apartment block, which is occupied by many families. These families are as separated from each other as are the occupants of a man-made block of apartments. There are mice too that carve out a number of nurseries in an old log.

Nocturnal mammals are probably the most active burrowers, but apart from the galloping animals, there are probably no orders that do not have some representatives that inhabit burrows. Even the female polar bear [89] digs a tunnel ten feet deep and bears her cubs in one of two chambers she fashions at the end of it. Since the cubs weigh only about a pound at birth and are born in early winter, the burrow is probably for their protection rather than hers because no other animal would challenge her, although the other Arctic predators would always relish a pound of tender bear meat in the land of scarce food. She stays with the cubs until they can hunt independently.

Other bears hibernate in burrows or dens that are often 8,000 or 9,000 feet above sea level. They are lined with moss and lichen, and dug with the entrance facing north, probably to get the best snow cover, because the cubs are born during the winter sleep. Such dens are like nests in that they are usually lined for warmth, and many animals go to great pains to do this. A few of the lesser known ones are the slender-tailed meerkat,[90] the nine-banded armadillo,[91] the paca,[92] and the raccoon dog of east-

25. The house-building rat, *Coniturus conditor*, builds a community apartment block in which many families live without the apartments encroaching on each other.

26. These mice manage to carve out nurseries in old logs.

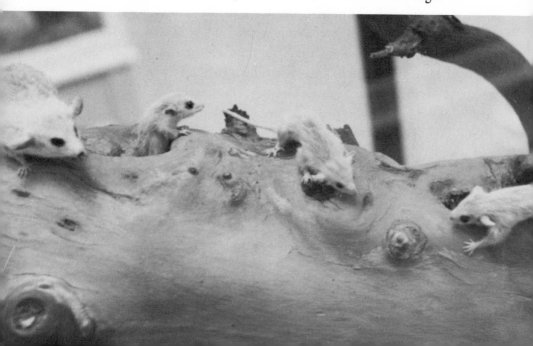

tern Asia,[93] which often shares a burrow with a badger or a fox. The *Mustela nivalis* weasel often takes over a rabbit's burrow, but it also uses a hollow tree or a hole in a wall, which it lines with leaves or grass. The tayra [94] does much the same thing, but this animal has a tendency to live in family groups, sometimes of large numbers.

Although the number of mammals that build truly birdlike nests is exceedingly small, some of those that do are quite well known. The female coati [95] isolates herself a week before the birth of her young and builds a tree nest, where she delivers and cares for them until they are independent. The female eastern cottontail rabbit [96] digs a shallow nest about six inches deep and lines it with dried grass and her own fur. The young are born there and she covers the whole thing with grass and leaves to camouflage it, returning to them only to feed them.

Even some primates do this. Toward the end of her pregnancy the female fashions a nest, often from her own fur, and delivers her young in it. The lesser mouse lemur [97] is an example; and the aye-aye [98] even constructs two nests, one to sleep in and the other for her family.

6. Giving birth—mammals

Birth can be a painful process for the human female, although it does not appear to be so distressing for women from less-advanced cultures, and this is perhaps because, like other mammals, they lead a more natural existence than their so-called civilized sisters, getting more exercise so that their bones and muscles are in better condition for childbirth.

Long before it is ready for birth, an infant is active, moving its position, and responding to muscle stimuli. In human pregnancy this occurs by the sixteenth week. In another two weeks the fetus' movements may extend to putting its thumb in its mouth, and by the twenty-fifth week its eyelids may raise and lower. By the thirtieth week it might even scratch itself or move its mouth as though yawning.

Movement increases as birth approaches, and ultimately there is an increase in the mother's uterine muscle spasms, which are normal in a pregnant female and which hasten the expulsion of the infant. This is the period known as "labor," and it may last from a few minutes in lower mammals to many hours in humans.

A headfirst birth position is essential in most mammals if long and difficult births are not to end fatally for the offspring. Any constriction of the umbilical cord would cut off the infant's blood and oxygen supply, and if it is unable to start breathing air at once, it will suffocate and be born dead. The very opposite position is necessary in marine mammals like whales and porpoises. The young of these animals are born under water, and if they were born headfirst, they could drown before reaching the surface to breathe; so they are born tailfirst.

Those marine mammals that deliver their young at sea frequently need aid, and the female porpoise almost always has a "midwife" assistant. She whistles in a particular manner, and an assisting female responds by helping her lift the baby porpoise to the surface to breathe. Whales also summon several members of the pack to help the mother push her newborn to the surface and support it there until it is breathing properly. Once breathing is well established, all other normal activities including suckling can take place under water. There is nothing strange about this. Even the baby hippopotamus suckles under water.

Most mammals lie down when about to give birth. This relieves the muscles that support the female's weight, so that she can concentrate on the business of labor. It also ensures that the infant will not be injured by falling to the ground. Exceptions to this habit of lying down have been reported, but they are rare.

Marsupials adopt a somewhat different posture. Their newborn are extremely small, immature, and delicate; and it is essential that they be given the opportunity to reach the pouch or the nipples as quickly as possible. Animals such as the wallaby, therefore, sit with their tail forward between the rear legs to produce the shortest possible route from the vagina to the pouch, which the mother cleans out just prior to the birth.

Many bats hang downward by the hind feet when resting, and this is a much more convenient birth position than it may appear to be. A female can deliver her young straight onto the breast. She has no difficulty in cutting the umbilical cord with her teeth before eating the placenta, which may be her only food until she can hunt again. Some species deliver into the membrane that stretches between the hind legs. The infant then clings to the fur and is helped toward its mother's nipple. There have been reports of female bats giving birth while in flight, but this would be difficult to substantiate and seems doubtful.

When a mammal is about to be born, the pressure from the mother's uterine contractions eventually ruptures the membranes that seal it off from the cervix and the vagina. Sometimes, however, if this membrane protrudes along with the infant before breaking, the mother breaks it with her teeth. If breakage occurs normally, the infant's head then passes into the cervical canal to reach the vagina and the outside world.

When the infant is first released from its maternal sanctuary, it is still attached to its umbilical cord and connected by this to the placenta; but immediately after birth the umbilical arteries contract and close to prevent any blood from flowing back to the placenta from the infant. The veins, on the other hand, remain open long enough for some of the blood remaining in the placenta to drain into the infant, then these veins too contract.

The cord breaks spontaneously in many mammals, but the mother quickly bites through it if it doesn't. It is only possible to speculate on a reason for the mother's eating the placenta. Nourishment is one possibility, and so is firm identification of the infant by odor. It may also ensure that no trace is left to bring predators to the area by scent.

The mother's next task is to lick the infant clean. This is such an ingrained habit that some isolated human groups still do it. Certainly, remote Eskimos are reported to do so. Animals kept in complete isolation, so that they have never witnessed birth of any kind, still do this automatically, which is interesting because animals reared in isolation do not always do other things as automatically as they do this. Some are even unable to copulate successfully when they have had no example.

Some of the land mammals assist birth too. Not only does the mother herself help the birth process, but also the father or other members of the group. Baboons have been known to help draw the child from the vagina and, after cleaning it and eating the placenta, place it at the breast so that it will find nourishment immediately. Desmond Morris has reported that a female orangutan,[99] although isolated from other orangutans, used her own hands to help deliver her infant, and then gently breathed into its mouth to ensure its respiration. The male golden lion marmoset [100] has been known to aid in delivering his mate's infants, and even to adopt the task of cleaning them after birth.

Although we are dealing with mammals, it can be mentioned in passing that the instinct to aid in the delivery of offspring may be deeply rooted far back in the animal world's ancestry, even though it is obvious in only a few nonmammalian species. The female alligator lies over her nest as the time approaches for her eggs to hatch. When she hears this happening and the infants calling, she digs away the earth covering and releases them. Then she leads them down to the water. The crocodile often does the same.

7. Being parents

Although countless animals spawn and pass on, never knowing their offspring, and even becoming cannibals if they encounter them by chance, there is a great number of animals that care for their young until they are mature and well trained to move off alone into the dangers of their wild world. From one kind of parent to another there are many levels of parental care, and although this care reaches its peak in birds and mammals, surprising examples are found among fish and amphibians as well. We could even say it is found among invertebrates if we consider the bees and ants that guard, feed, and carry their offspring as well as their eggs away from danger.

EDUCATING THE YOUNG

How important is parental training? There are experts who dispute the possibility of such a thing among animals other than humans; they suggest that there is only learning, not teaching. But many newborn animals are so helpless that they would not even find their mother's breast were she not to place it where the odor starts up a reflex mechanism in the infant, prompting it to open its mouth and take hold. Both monotremes and some marsupials assist the young to the mammary glands or nipples. In those species without pouches the young just remain suspended from the nipples and hair as long as they are inactive.

Parental and family example seems essential to many young animals. Both action and feeling seem to be stimulated by example. Without it a kitten and a young rat can be reared in harmony; the weasel and the rabbit can learn to live together; or the wolf cub and the lamb, so long as they are isolated from others of their kind and fed. Aggressiveness is augmented by example, and it is by example that young animals learn to create their own "safety perimeter." This is the distance that must exist between an animal and an enemy or other animal in order for the first animal to feel safe. If an enemy comes closer than that distance, it becomes necessary for the animal to change from passiveness or retreat and to attack.

The young wallaby seeing a python for the first time would never be aware of the fact that the snake could kill it, but by seeing and feeling its mother's reactions to everything that moves, it learns what is dangerous. Right from the beginning she communicates her feelings to the infant, which is just as well because there is seldom time to learn about such situations by experience. Just one encounter with a predator can be fatal for an infant. So parental training seems important to the higher animals, and in many species there is even a system of caring for and training orphans.

FISH

Few fish train their young for what is ahead of them, even though many guard and protect them until they can hide and fend for themselves. Even this short association with adults, which may last no more than a few days, may be enough to give the fry a groundwork of self-preserving behavior. In Chapter 1 we encountered the mouthbrooders, which not only hold their eggs in their mouths until they hatch, but also take the young into their mouths when there is danger. The young of the gaff-topsail catfish [101] do not leave the parent's mouth entirely and take care of themselves until they are four inches long, and one assumes that by that time they must have benefited by example.

One mouthbrooder, *Haplochromis multicolor*, provides a good example of learning by mimicry. The female looks after the young for a period of about ten days, taking them back into her mouth when there is danger. As she begins to pick up the little fish by sucking them into her mouth, all the rest rush to her mouth in an effort to get in. Before long they will even go to a model fish or a dead one as well as to a live one. Nor do they seem able to differentiate between females when heading for a mouth. They just bolt for cover.

Both male and female black bullhead catfish [102] seem to guide their school of fry in what might be educational sessions, and one wonders why this is not seen in more species. But most fish fry begin to school by associating in pairs that gravitate toward other pairs and form groups. If adults do not provide an example for this, it may be related to the need to head into a slight current. After the male stickleback has partially dismantled his nest, leaving just the foundation as a kind of resting site for his fry, he begins to round them into a small shoal, and once they have learned to stay together, he leaves them.

BIRDS

It is among the birds that we first see the pair bond. Ordinarily, this union of a male and a female is not very important, but during the breeding season it is. It enables the male to go off hunting without being concerned about the female taking other males while he is away. It reduces friction within groups, and is a definite contribution toward survival. This is our first glimpse of real family organization in caring for and protecting the young as well as jointly incubating and guarding the eggs.

Not only do birds expend a great deal of energy in searching for food for their chicks, but many of them clean the nests, picking up pieces of broken shell and chick's droppings, and disposing of them. The lyrebird feeds its single chick about every half hour throughout the day for up to six weeks, or longer if it is reared in a tree nest. After every second or third feed the chick presents its rear end to the hen and excretes in a gelatinous covering, which the mother carefully takes in her beak to a pool and drops into the water, or, if there is no water near, buries in soft soil. Thus no trace is left to alert predators.

The rearing patterns of some birds include some very specialized functions. Both male and female doves feed their young with a milky secretion from their crops until the chicks are old enough to take more solid food. Surprisingly, this milk has seven times more fat in it than cow's milk, but little seems to be known about how the milk is manufactured in the crop.

Not all birds are helpless at birth. Many are ready to fend for themselves from the moment they break out of the egg. So we see two very different kinds of newborn. Chicks that are born naked and helpless require a safe nest and continuous attention. They raise their heads when they detect an adult nearby, open their mouths wide to show their highly colored gapes, and clamor loudly for food. The parents are apparently stimulated by the sight of a gape and feed the largest one, the one that can reach out the farthest to them, or sometimes the one with the brightest throat color.

Because parent birds seem quite unable to differentiate between their offspring, not every chick gets its needed share of food. The quickest and most vigorous gape gets it all until its owner is satisfied. Then the frenzy slows down, and the other chicks get more attention. When all the young are satisfied, the parents themselves feed. But when there is only one chick, and the species is a colony breeder, both parents and chick can recognize each other by their sounds, even when the noise of countless thousands of birds around is deafening. The laughing gull [103] chick begs its food, not by opening its gape, but by pecking at its parents' dark red bills, which seem to attract it in much the same way as the colored gapes of chicks attract their particular parents.

Young birds grow fast. They stand at the edge of the nest and flap their wings to test them, ultimately fluttering off. They must still be fed, however. In fact, one wonders if some chicks would ever bother to feed themselves if parents did not guide them to places where food is obvious, and then gradually leave them to try picking it up as they have seen the adults do. Some terns place food in a chick's mouth for the first few days of its life, but later drop it as they hover above, so the chicks simply must learn to pick it up or starve.

It is a relatively long, hard life for some bird parents. An eagle in Kenya, *Stephanoaetus coronatus*, breeds only every other year, because the young do not complete the postfledgling period for up to 350 days. The male hornbill feeds both its female and young for two months. They are walled up in a hollow tree, as described in Chapter 5, so are entirely dependent on the male for everything. If he is killed, some other unattached male takes on his responsibilities in providing for the widow and orphans.

Seabirds, which usually raise only one chick at a time, often leave it for long periods, which compels it to learn to fend for itself. The Manx shearwater [104] will desert the nest and go off on food-gathering expeditions that last up to eight or nine days, although in the initial feeding period chicks are fed almost every day.

Sometimes these long absences are also related to a shortage of fish in the area. Gannets, which nest in huge colonies, can satisfy

27. Seabirds like this sooty tern often have to fly great distances to find food, because they breed in such large colonies that the adjacent areas soon get fished clean.

their family needs within a radius of one hundred miles in the early stages of rearing their young, but since the latter take about ninety days to mature, it may become necessary to fly four hundred miles or more for food as the area becomes fished out. Other birds appear to teach their young to fend for themselves by systematic neglect, or perhaps it is only loss of interest after the exhausting period of rearing.

Jackass penguins [88] have an interesting habit of assembling what one might call their adolescent chicks into large schools, which are then tended by the whole colony in a communal effort until the young are old enough to be shown how to fish alone. Then they form their own fishing groups, which may stay together for three or four years.

Most birds keep their chicks warm at night by covering them

28. The magpie goose shows her chicks where food is and how to pick it up, even though they are advanced enough at birth to feed themselves.

with their wings and bodies, and the parents must often protect their chicks during the day also, because rain and excessive heat or cold will kill them. In addition to this, concealment, the instinct of chicks to "freeze" at any sign of danger, and aggressive defense by parents contribute to the survival of these helpless little creatures.

Chicks of species such as ducks, fowl, plover, and prairie hens, which are born complete with down, are able to walk as soon as they have dried out. They can feed themselves also, but even though they are able to peck, they must first be guided to food and stimulated by a parent's example. They learn and mature very fast, and some of the mound birds can fly on the third day. Usually, those birds that are so advanced at birth have had a long period of incubation.

Showing them where to find food may not be all that birds contribute to their chicks' training. The female satin bowerbird mimics the sounds of the predatory birds in the area, and it does seem as though she does this more when she has young chicks, as though she were giving them a course in caution. But this assumption may be too simple, and it may be quite unjustified.

When monogamy does occur in nature, it seems to be a partnership only for animals that never leave a particular piece of territory or that need almost a year to raise a brood. Some of these monogamists are geese, swans, owls, eagles, parrots, parrakeets, and ravens. Two that appear to stay together for life are the Canada goose [105] and the bald eagle.[106]

MAMMALS

It is among the more than 4,000 species of mammals that the longest periods of parental care and training are encountered. In man it seems to be growing longer, up to twenty-five years or more among the more affluent sections of society, which is a certain indication of man's fast-evolving superiority. It is difficult, however, to know just where to begin describing the rest of the mammals, because they are so varied and interesting.

Being born, like giving birth, is one of the most complicated episodes in all of life, and yet in raw and savage nature it must

29. A mother Barbary sheep stays close to her newborn offspring, waiting for it to gain the strength to move off with her.

often be accomplished with speed if it cannot be done in safe hiding. There are always predators waiting to snap up a tender newborn animal. Rapid birth is especially necessary when animals live in tight groups that are constantly on the move. A mother giving birth must fall behind the group, and she then becomes very vulnerable.

Not only must the birth be rapid in such species, but the newborn animal must be active almost immediately, so that it can escape with its mother to the relative safety of the group. We can see this precocious activity at any time in familiar animals in parks and zoos. Gazelles can run when they are an hour old, and they need to, as do most other similar animals, because vultures, jackals, wild dogs, hyenas, cheetahs, lions, leopards, baboons, and a host of others will take them if they do not. Young zebras are the same. They stand almost at once, can walk within minutes, begin to trot very soon, and then gallop with their mothers as they rejoin the herd.

Young animals must also be able to identify their mothers. A female establishes her offspring's identity by licking it clean when it is born, but the infant itself must use as many other cues as possible because of its inexperience. Odor must play a part, but so does vision. The first thing its eyes see becomes identified as mother, especially if she is moving, and unless an unfortunate accident occurs, mother will always be standing over it in its first conscious moments.

Inevitably, some newborn animals are lost, but nature makes provision against the loss of too many at one time. Mating takes place in the whole herd at about the same time, so many infants are born almost simultaneously. They come so fast that it is quite impossible, under ordinary circumstances, for all of them to be killed.

The richness of a mammalian mother's milk is a vital factor

30. The lamb looks up for the first time, and the first thing it sees is its mother. At that moment she becomes indelibly identified in its mind.

in the speed of an infant's growth. Marine mammals appear to have the richest milk of all, and the harp seal pup [107] doubles its weight in five days as a result of this. The southern elephant seal pup[108] increases its weight from about 112 pounds to 224 pounds in eleven days. The California gray whale pup gains 200 pounds a day. A high protein content in the milk evidently contributes to this rapid growth. The galago [109] is similarly endowed.

The female porpoise has special muscles near her nipples which force milk into the baby's mouth at a great rate, enabling it to take in considerable quantities in the short period between surfacings for air. The female gray whale can expel thirty gallons at a time in this way.

The female midwife that attends the porpoise mother while she is giving birth becomes a kind of foster mother, staying to help the mother guard the infant and to baby-sit for her. In a number of mammalian species juveniles are even collected into large groups and supervised or cared for by one or two adults. African wild dogs, *Lycaon pictus*, have communal nurseries of thirty to forty pups cared for by one female while the rest go off hunting. The mongoose [110] also organizes things so that one adult stays with a number of young while their parents hunt, and so do lions.

31. Finding food appears to be no problem, even without experience.

When the female Pribilof seal gives birth, the pup is able to move around at once. The mother calls to it and sniffs it. It answers her, and the sounds and odors identify them to each other. After the first few days mothers do not protect their pups, they only feed them with their highly concentrated milk and then go off to feed themselves, often staying away for five to eight days at a time. The bulls do not leave the area even to feed, and their constant aggression into neighboring harems makes life hazardous for the pups, which are liable to be crushed under the heavy bodies of the bulls. So the pups tend to gather in groups of their own. They seem to receive no help from their parents in learning, and they learn to swim by playing at the water's edge in their own groups.

Many mammalian mothers protect their young by hiding them, visiting them only to feed and tend them. The European hare [111] disperses her young to separate places soon after birth, and then for about four weeks visits them one by one to feed them. This way, if a predator finds one of them, the others usually escape. The female Chinese water deer [112] is another animal that hides her three or more young in separate places, and visits them to suckle and tend them.

The young of a tree shrew, *Tupaia glis*, are hidden in a separate nest, which the male constructs just before their birth, and then they are visited only by the female once every forty-eight

hours. She suckles and cleans them and then leaves them, returning to the parental sleeping quarters somewhere else. Even the
echidna [57] operates in much the same way. When the young
one's spines develop, she leaves it in a dry shady place and
returns to feed it until it can fend for itself.

Female cheetahs keep their young with them and train them
for two years. The female grizzly bear [113] feeds her three or
four cubs for up to two years. If we consider that they weigh
only 18 ounces at birth, soon reach 100 pounds, and when mature may weigh 1,500 pounds, this is not surprising. That the
offspring of some animals may require training by parents or
learning under supervision has been supported by experiments
with rats. Male rats that are raised in complete isolation after
weaning are quite disoriented and unable to achieve normal relations with other rats. They cannot even copulate with normal
females.

32. Solicitous giraffes stand over one of their infants.

The koala [48] carries her infant on her back for six to twelve
months, after having carried it around in her pouch for a similar
period. Frequently, by the time the young one goes onto her
back, another newborn is ready for the pouch; and in rare
cases she may have one in her pouch and two on her back
at different stages of growth. This may complicate her life,
because when a young one is six months old, it must be weaned.
Before it becomes leaf-eating, there is a transition period of a
month during which she provides a soup of digested leaves from
her anus, which mysteriously at that time is quite clean.

Having offspring of different ages to care for is not uncom-
mon in marsupials, and nature has provided for this by making it
possible for two different forms of milk to be available in two
separate nipples.

Many other animals carry their young on their backs—lemurs, elephant shrews, sloths, opossums, ant bears, and scaly
anteaters, to name but a few. Many female fruit bats carry their
young until they are weaned, five to eight weeks after birth,
and these young usually cling to the breast fur. In some species,
however, when the young one is a few days old, the mother
carefully hangs it up by its wing claws and leaves it so that she
can hunt for food more easily. In other species a female
that has borne young releases an oily secretion from glands
around her nose, and this secretion stains her chest fur, giving off
a strong odor, which identifies her to her offspring—something
that is common to many mammals and is even encountered
among the primates.

Some flying foxes,[114] or fruit bats, keep different camps for
summer and winter, and the young are born in spring in the
summer camp. But in the autumn, when the adults disperse
throughout their normal range, the young move in packs to the
winter camps, congregating from many directions, and there

they learn among themselves all the group habits they did not acquire with their parents.

Obviously, many animals have infinite patience in rearing their young. Not only that, but most of them tolerate endless teasing and play from them without becoming bad-tempered. We see this all the time in lions and other cats, wolves, dogs, and foxes.

Some mammals are driven away or leave their parents soon after they are weaned, and these take up an independent life. Others stay with the family and/or tribal group or herd—some only until they are mature, others permanently. Many animals that live in herds segregate themselves, the males in one herd, and the females in another, where they keep their young and help them develop the behavior habits of their species. Bison, wild pigs, and deer are examples of this.

PRIMATES

One of our closest relatives, the chimpanzee, provides us with a most interesting study of family life. At birth, the infant has hair on its scalp only, like most human children, and it is just as helpless. But it does have considerable strength in its hands and feet, and with these it hangs on to its mother, hardly letting go for the first four months.

After this time it may venture a few feet from her, but the mother reaches out quickly if it loses balance. When it is a year old, the young one begins to play, and at two years it is incessantly active; but it still sleeps with its mother and will continue to do so for at least another year.

Chimpanzee parents are very gentle and tolerant, and the whole family group is very playful. At eight years the young one has reached puberty, and after another three or four years of adolescence it reaches adulthood. Theirs is a very devoted family life in every way. They attend to each other's injuries,

33. Chimpanzees, *Pan troglodytes*, form most affectionate family groups, and the young are protected with great care. A baby chimpanzee usually does not let go of its mother's fur for the first four months of its life.

34. Like their cousins the chimpanzees, orangutans form close and affectionate family groups. They show just as much parental affection as humans.

remove thorns, remove grit and debris from the eye for each other; and the males even share their females without quarreling.

Like the human mother, the female chimpanzee usually holds her baby with its head to her left side. It is said that in the womb before its birth, any baby's system is somehow attuned to its mother's pulse, and when later it is nursed by her, having its head on her left side, where the heartbeat is strongest, is soothing to it. It gets a sense of security and relaxes into easy sleep.

When a young male chimpanzee reaches adolescence, the mature females teach it to be aware of its genital organs and show it how to copulate—perhaps the only animal to carry education to this point.

Our other close relative, the orangutan, is perhaps less social and remains more often in single family groups or small mixed groups. It shows just as much gentleness and concern for its offspring as the chimpanzee, and in fact there is a great similarity in the parental behavior of the two animals. The gorilla also lives in family groups, which can be of considerable size, and it too exercises great protective behavior toward its young. Probably all the anthropoid apes encourage their young to walk by holding them up by both hands, just like a human mother.

In addition to the great apes, many of the smaller primates show equal parental concern. Perhaps the most study has been given to baboons. Within a few hours of birth, the infant baboon is able to hang on to its mother's fur with a minimum of assistance. At three weeks of age it requires no help at all. Grooming by the female is intense and constant, and she gives her infant her undivided attention. The males of the troop collectively protect mothers and infants, and this protection extends to the young after they have left their mother's sides at six months of age and more.

35. The female tree kangaroo provides a safe retreat for her young one until long after it is able to get around itself.

Both tree-dwelling and ground-dwelling primates create this secure and protected environment for the young in a troop. In the latter, however, the troops are larger, and protection, therefore, is greater in certain circumstances. As the young grow and mature, the adults encourage them to join in social contacts with other young, and they are protected only in times of serious danger. Ultimately, the adults may drive them away to some extent to encourage their independence.

A young baboon usually has no brothers and sisters, the last offspring of its mother being two years old and independent. But there are many other infants of its own age, because so many births are simultaneous in the wild. One-year-old infants become the first serious playmates for the offspring when it initially leaves its mother's side at about nine weeks of age.

The males permit almost any liberty from the young; certainly they tolerate things they would not tolerate from older animals, and they leap to the infant's aid if it is involved in unnecessarily rough play. After the infant is nine or ten months old, the female prevents her infant from suckling and riding on her; it is too heavy and too active anyway, but, more important, she is moving into breeding condition again.

Rhesus macaques [115] show the same protective habits as the baboons, the males enabling the females to concentrate all their attention on the infants. The pigmy marmoset's [116] behavior is rather the reverse. The litter, which includes two, is carried by the father, one on each thigh, and returned to the mother for feeding only. The male golden lion marmoset [100] also takes over the young for six or seven weeks after birth and returns them to the female for feeding only. Later he also prechews food for them, but his care wanes after about fifteen weeks. Tamarins [117] and titi monkeys [50] are just as unusual in that the males rear the infants until they are weaned, the females only feeding them, but after the infants are weaned both parents share the task of rearing.

36. The female koala continues to carry her young one around even with another in the pouch, and will in fact sometimes have two on her back of different ages.

37. The Australian opossum also carries her young on her back, but not often does she have more than one at a time.

38. The American opossum may have several young on her back, but these opossums have prehensile tails, which the young are able to twine these around their mother's. So they travel with comparative security.

39. Many female fruit bats carry their young against the breast for up to eight weeks, but other species hang their young up by the claws and leave them while hunting for food.

The Barbary ape [118] enjoys sharing the group's infants. All the adults are attracted to all the infants in a group and pay them a great deal of attention. The dominant males are particularly attentive during the first twelve weeks, and the whole social structure is tighter than in the rhesus.

Although lemurs are lower primates, their parental habits are in many ways similar to those of the higher monkeys and apes. In the sifaka,[119] which lives in troops, the single young one clings to its mother's abdominal fur for three months, and then rides on her back for two or three months, by which time it is half grown. Sifakas have a close family and troop relationship, and they all want to groom any newborn infant, just as humans crowd around a maternity ward window. By grooming the newborn, they set an example in grooming for the young. Such familiarity is not permitted by the female ring-tailed lemur.[60] The mothers of this species permit only other mothers to groom their newborn.

40. These two tiny rheas are orphans, and have been adopted by their uncles, who will care for them as long as they need it.

THE CARE OF ORPHANS

In a number of species throughout the mammalian class, motherless animals are cared for by other females or even by the tribal group. We see this behavior in birds too, but we also see a great deal of variation in the degree to which this substitute rearing is successful. Among some birds, for instance, those that are reared by foster parents do not always recognize the warnings of these parents, and they are taken by predators.

Sometimes adoption is only possible by a mother that has lost her own offspring a few hours before. Ewes do not ordinarily feed motherless lambs, but when a female receives a motherless lamb early in her own period of loss, she usually licks and adopts it. Licking it for twenty minutes or so seems to give it distinction from other lambs and so establishes a normal mother-child bond.

Adoption does not always have to be made by an experienced mother. Animals with strong social groups adopt orphan infants either individually or communally. Female elephants do this quite readily. It is enough that the adults themselves have had a normal rearing and have learned by example. Virgin female rats, and even males that have been exposed to newborn young within the group, often care for orphan rats. There are, in fact, many mammals that do this, and strangely, in some species the males may be more likely to adopt orphans than the females are.

Male wild dogs often feed orphaned pups, and if a female grizzly bear is killed and leaves cubs, these are invariably cared for through adoption. The restless cavy [120] infants seem to enjoy the same social security, even being suckled by foster mothers. Lionesses too suckle orphaned cubs. Apes, baboons, and many monkeys defend unrelated and defenseless young of their species, and so do porpoises.

A young animal that is old enough to scrounge enough food to survive may not be adopted, even though its training is incomplete. Often such an animal turns out to be an unsuccessful breeder. It is not only unaware of the necessary physical actions, but it relates rather badly to the opposite sex in general. Orphaned females are usually poor mothers too. They never seem to develop the close and affectionate ties to their young that they themselves were deprived of. We also see this quite clearly in many humans.

The more complete the social isolation of these animals is, the worse the effect is on them and the less successful they are in creating a good relationship with their own offspring and with other members of the group. This applies much less to lower orders; in those, sexual behavior is usually less complex. But even in these animals, experience and example seem to prepare them for sexual activity at an earlier date.

PARASITIC BIRDS

Sometimes birds become foster parents without volunteering for this task, other birds depositing eggs in their nests and then leaving. There are few places where some kind of cuckoo or cowbird does not exist, a seemingly carefree member of the otherwise industrious bird world that refuses to accept the normal responsibilities of parenthood. These birds leave their eggs for other birds to incubate, and these other birds also rear the chicks, which often become so much larger than their foster parents (one cuckoo reaches a length of twenty-four inches) that the latter must stand on the chick's head to feed it.

Some cuckoos lay an egg in another bird's nest and at the same time remove one of the eggs already there. Since the cuckoo egg frequently resembles those among which it is placed in both color and size, the deception usually succeeds. If this is done to a dozen nests in an area, the cuckoo population can be as much of a menace to other birds as the predators are. It can be even worse when, as in some species, the cuckoo lays four or five eggs in each nest.

The cuckoo's incubation period is often a little shorter than that of the host bird, so that the cuckoo chick gets a head start in feeding. In some species it will lever the other eggs out of the nest, or do this with the fledglings when they are born. Even if it does not do this, it is likely to be larger than the others, and its greater gape gets all the food, the smaller fledglings starving. So the removal of an egg from the host's nest is not all the damage that the cuckoo does to the foster family.

There are a few birds that lay their eggs in another's nest when they cannot get a site or nesting material for themselves, but this is not a systematic parasitism like that of the cuckoos.

Part III

Controls
on Heredity
and Behavior

8. Inherited characteristics

Chromosomes, which are contained within the ova and sperms, contain the genes that are the basic controls for all of an animal's characteristics. Animals do not all have the same number of chromosomes; for example, man has 48, the horse 60, the ascaris 4, the crayfish 200, and one radiolarian has more than 1,600. We do not know how many genes there are in a single chromosome, but it is a vast number. All we can say is that each sperm and ovum contains thousands, perhaps millions, of genes, and billions of combinations are therefore possible.

The possible combinations of chromosomes alone is fantastic. In man the 48 chromosomes in each sperm and ovum can produce any one of 281,474,976,710,656 combinations. Even if two individuals somehow managed to obtain the same combination, however, this would not make them alike, because within each chromosome the countless genes also form their different combinations. Except for monozygotic twins, therefore, no two animals can be identical.

Genes

Genes are complex protein factors containing what has been called a "code," the pattern of the features and characteristics of each individual animal. It is a great collection of information, all

129

contained in a single chemical called "DNA," which is an abbreviation for deoxyribonucleic acid. These genes control every characteristic and predisposition—bone size, skin color, height, kind of nails, facial shape, tongue sensitivity, disease tendency, ear shape, courage, aggressiveness, intelligence, fertility, special abilities, and even the actual patterns of behavior that can be learned or the ease with which they can be learned. Every conceivable quality in an animal is genetic, even though it can be affected by environment.

The genes in the chromosomes are reduplicated every time a cell divides, and so they pass to all the cells of the developing body, and they are passed to all following generations. Since mutant genes are relatively rare, animals evolve into different types and forms only over long periods, often millions of years. In the process of evolution changes in temperature, food, soil chemistry, and other environmental factors can make it impossible for some genes to duplicate themselves. The genetic combinations then produced when an animal breeds may or may not survive. So environment and heredity are interacting factors in evolution.

Since the number of gene combinations is so incalculable, the loss of even large numbers of them because of changing conditions would not necessarily affect the basic demands of a living organism. Mutations, therefore, may be either this loss of genes or a rearrangement or even a reduplication of the genes within the chromosomes. Any rearrangement within the chromosome will, in fact, produce a changed chromosome.

CHROMOSOMES CONTROLLING SEX

In all sexed animals there are at least two chromosomes that determine the sex of the fertilized ovum. These are known as X and Y chromosomes. In mammals an XY combination produces

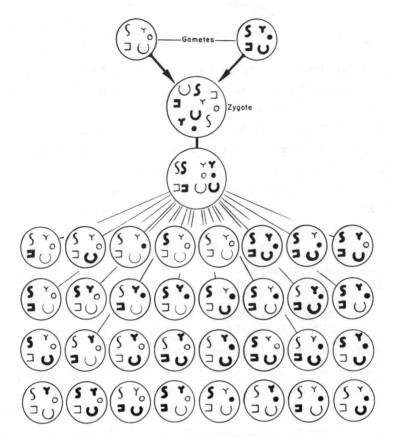

41. This diagram shows that gametes from an animal species with only five chromosomes can produce thirty-two different chromosome combinations. There are countless genes within these chromosomes, so the number of genetic combinations is fantastic.

a male; an XX combination produces a female. So if a fertilized ovum could be examined with a microscope, it should tell us what sex an offspring would be. Birds have exactly the opposite pattern from mammals. An XY combination produces females, and an XX produces males.

Color Inheritance

Quite obviously, when we try to illustrate the principles of heredity and the way in which certain characteristics are passed on or become dominant, we have a vast fund of possibilities from which to draw, but it seems best to confine ourselves to familiar situations seen in everyday life. One of the most obvious of these is the color of the eyes and hair, and so for the first example we will use human characteristics, although the same principles apply to animals of other kinds.

When two people with eyes of the same color have children, these children frequently have similar eyes, either all blue or all brown, but because brown is a dominant color and blue is recessive, two brown-eyed people may also produce a percentage of blue-eyed children. When one parent has brown eyes and the other has blue, the children will also be mixed, some with brown and some with blue eyes. There are many shades of brown and many of blue, all of which result from different genetic combinations and from the incomplete dominance of some of them. The hereditary patterns for hair color and skin pigmentation are similar, light hair and skin being recessive. Every other human characteristic, anatomical or physiological, and in some cases even disease, is regulated by an inheritance factor, which may be either dominant or recessive.

The simple and easily observed feature of hair color is illustrated in Figure 44. There are naturally more combinations of color genes than those that produce black or white or patched patterns, and among most animals, coloring is not that simple. It conforms to camouflage patterns and colors that are essential for survival. Those animals that do not inherit the necessary pattern seldom survive long enough to pass on their difference to a significant number of offspring. The most obvious example of this is probably albinism, which is a scarcity or absence of pigmentation in the skin, hair, and eyes.

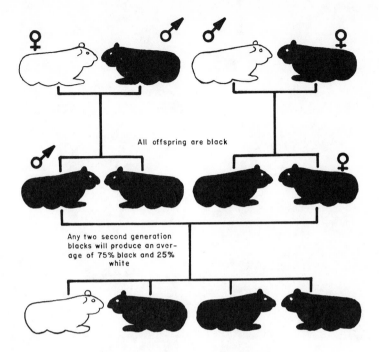

All offspring are black

Any two second generation blacks will produce an average of 75% black and 25% white

42. When a black guinea pig mates with a white one, more black than white offspring are born, because the light color is due to a recessive gene; but the recessive white may reappear when two of the offspring pigs are mated.

ALBINISM

Albinism is not rare, but it is seldom encountered in animals because albinos do not survive for long in the wild state. Not only are they very easily seen by predators, even at night, but often their own species destroys them. They draw attention to the group and thus jeopardize its safety. So at the very least they are avoided by their own kind. Also an albino's eyes are almost always defective, and this adds to the odds against survival.

Albinism is a gene defect that creates a deficiency of the enzyme that produces pigment. An enzyme is a complex protein having the power to accelerate chemical reactions in the body processes, and when the appropriate enzyme is deficient, the

43. These wallaroos are pure albinos, their eyes and noses being quite devoid of pigment.

44. Albino deers are not unusual, but they do not often survive in the wild state.

45. This beautiful male peacock is so albinic that his tail does not even show a pattern of figuring.

46. An albino crow looks odd beside its normal companion. Even its legs are colorless.

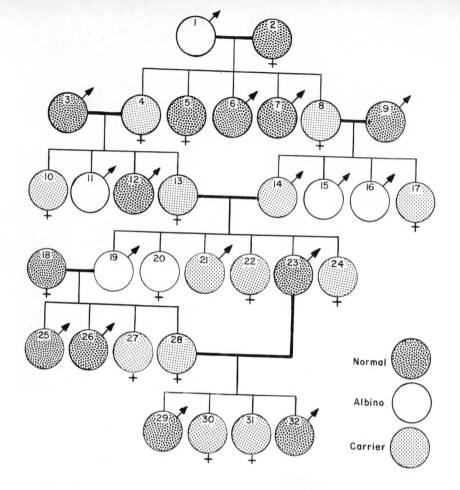

Normal

Albino

Carrier

47. In this diagram, an albino male (1) mates with a normal female (2). Their offspring consist of two female carriers (4 and 8), two normal males (6 and 7), and one normal female (5). Both carrier females mate with normal males (3 and 9), and both these pairs produce albino males. Numbers 3 and 4 also produce a normal male (12) and two female carriers (10 and 13), and numbers 8 and 9 produce a female carrier (17) and a male carrier (14). Numbers 13 and 14 are cousins, and when they mate, they produce more albinos (19 and 20) and carriers (21, 22, and 24) than normal offspring (23).

When one of these male albinos (19) mates with a normal female (18), the result is much the same as when 1 and 2 mated. One of the female carriers (28) mated with her normal uncle (23), and the result is the same as when 18 and 19 mated. These results are all possible, but so are numerous others. Female carriers of the third generation (and we can assume that number 8 was of the third generation) can produce male as well as female carriers (14 and 17).

formation of pigment never takes place. So the animal's normal body and fur pigment are absent, and the irises of its eyes appear pink, because the blood in the blood vessels shows through.

Obviously, the hereditary factor is not a simple one in albinos. It is not the nontransmission of a gene for color, but rather the nontransmission of one for a chemical that influences the formation of pigment throughout the entire body. When two animals that have such a defect lying dormant in their chromosomes mate and produce offspring, the two defects combine and dominate the hereditary pattern so far as color is concerned, and the offspring are born completely albinic.

Apparently, when male albinos do live long enough to breed, they do not transmit their defect to male offspring. But female offspring carry the defective gene in a recessive form to the next generation, and some of the males of the third generation are albinic. When both the male and female have a recessive gene for albinism, they can have both male and female albino offspring.

There are also degrees of partial albinism. It may be restricted to the eyes alone, in which case it is frequently confined to males. In rare instances, the lack of pigment may be confined to one eye only, which can be seen in man and domestic animals. Or there may be albinic, or white, patches on the skin.

Not all white animals are albinos, however. Some, such as the Arctic fox and wolf, which live where the winter is long and snow abounds for much of the year, take on a white winter coat. The stoat, or ermine, living in Arctic regions, develops a white winter coat that is much sought after. Others, like the polar bear, have a permanent white coat. None of these are necessarily albinos.

Numerous animals, such as the white swan and the white heron, have a white color that is related neither to polar conditions nor to albinism. Like those animals with Arctic coats, which

48. In this rare albino trout, the iris of the eye is quite devoid of pigment, but the pupil shows black because it is merely a hole in the iris and there is no light behind. In the wild state this fish would not have survived because of its visibility.

show pigment in their nose, lips, and eyes, these birds have colored eyes and bills.

So if one sees two white animals of the same species, but one has color in its irises and the other does not, the former is not a true albino; the latter is. The one with pigment in the iris may, however, be a partial albino.

Some kinds of inherited characteristics are not always as rigid as those just described. Some of them are so well disguised by other characteristics that they can only be brought into useful effect by environmental influences. And in vertebrates, at least, an animal's hormonal complex is likely to affect the way it uses its inherited characteristics. Learning how to live effectively at an early age is even more important. When an animal's genetic characteristics are not of the very best, early learning may help to improve its chances of survival; if the animal has inherited all the best survival qualities—strength, speed, intelligence, aggressiveness, correct color for camouflage, etc.—then early learning may not be so important.

49. The Arctic wolf takes on a white coat, which blends with the winter snows, but it is not an albino. Its nose and lips are well pigmented, and its eyes are a golden brown.

9. The effects of environment on reproduction

People who are crowded together in congested cities are greatly affected by the pressures resulting from their confining environment. They are seldom so patient, gracious, or benevolent toward each other as those living in less-crowded conditions. The pressures of continuous close contact seem to breed more crime and violence, as well as more competitiveness and aggression. Animals also react to crowding and discomfort.

Experiments have shown that there is a point of crowding beyond which animals cease to live in harmony with each other and their surroundings. They become bad-tempered, aggressive, and violent. But crowding, or overpopulation, is only one environmental factor; there are others, such as scarcity or abundance of preferred foods, vagaries of climate, invasions of new predators, and disease that compound the crowding. However, the condition frequently rectifies itself in nature by the reduction of numbers through competition or migration, rather than by the compounding factors.

Many other things, such as unusual temperatures, overcast skies, unusual rain, snow, wind, drought, etc., affect the lives of animals, and some of these factors are very specific in their effects on breeding—especially in those animals that breed only once a year, with gonadal development being suppressed for the greater part of the year.

THE EFFECTS OF LIGHT ON BREEDING

The length of the day affects the development of the gonads in nontropical animals, and experiments in which the day length has been varied artificially have produced some interesting effects. When hamsters are subjected to a cycle with one hour of daylight and twenty-three hours of darkness, their gonads degenerate. Exactly the same thing happens if their eyes are removed, so daylight obviously affects breeding potential.

There is evidence to suggest that in some animals the light acts on the pineal gland, which, in turn, releases or withholds hormones affecting the gonads. Experiments with rats and foxes support this belief. In many other animals, however, the lengthening of the days stimulates the pituitary gland. This gland manufactures gonadotropins, which, in turn, cause the gonads to produce their own hormones. Although experiments have produced a great deal of knowledge, we still do not know whether the pineal or the pituitary gland is primarily responsible for the gonadotropic influence.

In at least some animals the gonadal response is not just to the length of the day, but to the *steadily increasing day length*. Those animals that breed in the autumn, so that the young are born in early spring, obviously have the very opposite response.

Experiments with birds have shown that they also respond to day length. The male Japanese quail [121] has a gland in the wall of the cloaca which secretes a foamy substance. This substance is transferred into the cloaca of the female during copulation and seems to be a necessary medium for the sperms. This gland, the production of the foam, and the testes are all affected when the day length is changed. In fact, the quail's whole reproductive system is affected. When the days are shortened, or castration takes place, the cloacal gland degenerates.

In these birds also the pineal gland is believed to be the intermediary factor, affecting both male testes and female ovaries. The effect of the pineal gland is very marked just before sexual maturity. Similar observations have been made in experiments with sparrows and other birds. No matter what the time of the year, lengthening the period of daylight encourages most birds to adopt incubation behavior, nest building, and molting. It can be seen at home by those who have canaries.

THE EFFECTS OF TEMPERATURE ON BREEDING

Although there can be no question of the importance of light in bringing animals to breeding condition, temperature is important too, because in some animals the lengthening days only affect the gonads if the temperature reaches a certain level. At least this appears to be the case with cold-blooded animals such as fish and reptiles.

Anolis carolinensis, an American lizard, only responds to the lengthening days if the surrounding air raises the body temperature to at least 32° C. (90° F.), and this is apparently so critical that the lizard fails to respond normally when the temperature goes above or below this. Apparently, the effects of temperature extend to embryonic development too, for the bones, especially the vertebrae and fin rays of fish, vary in number with the temperature at the time of incubation. This variation in the number of bones has also been noticed in *Ambystoma gracile*, which is a salamander.

There is a ground skink in the United States that shows a definite correlation between breeding activity and temperature. Around the Gulf of Mexico, where conditions are subtropical, this animal has a long breeding season with only brief and occasional periods of hibernation, so it is able to rear four broods of eggs in a year, sometimes more. But farther to the north this lizard's breeding season is shorter and its broods fewer. However, there are usually more eggs in each clutch, especially early in the season, compensating somewhat for the less frequent laying.

Very few cold-blooded animals can do anything about the effects of temperature variation on embryonic development, but pythons do seem to have some built-in control over this. The brooding female python, *Python molurus bivittatus*, not only coils around her eggs, a rare thing in snakes, but can regulate her body temperature by muscular movement. When the surrounding temperature drops below 33° C. (91° F.), spasmodic muscle contractions start along her body, rather like the shivering of birds and mammals, and this exercise raises her temperature, the amount of the increase being related to the frequency of the contractions.

In this way the female python can maintain a body temperature that is as much as 7.3° C. (13.1° F.) above the surrounding air temperature. Ordinarily, as the temperature drops, a reptile's consumption of oxygen drops also; its whole physiology slows down. But this does not occur in the brooding female python. Since she maintains her temperature by muscular contractions, she also maintains a normal metabolism and level of oxygen consumption.

In many fish, spawning condition is very closely related to water temperature, but some strange inconsistencies exist, even in two related species. For instance, the silver mullet [122] begins to school and spawn when the temperature rises in spring, but the striped mullet [1] moves to its spawning grounds and spawns when the water temperature drops. The temperatures that fish require for spawning are very critical and must usually be within a range of 5° C. (9° F.). The carp [123] is exceptional in having a range of about 10° C. (18° F.)—from 16° C. (61° F.), where activity ceases, to about 26° C. (79° F.), where it again tapers off. But except in the tropics, where day length changes little, the influence of temperature is shared in some way with that of day length.

There are other complicating factors related to temperature. Fish eggs can be affected while they are maturing if temperature changes take place before they are fertilized. The difference from year to year in the quantity of spawn in haddock [124] of the same size and age can be traced to the effects of temperature variations at certain critical periods in the development of the eggs. And even the growth of some adult fish, and as a result their fertility, is affected by such temperature variations. When the temperature drops unduly, the mortality rate among both adults and fry may be very high.

Another effect is seen in the common toad, *Bufo vulgaris*, which produces a much higher proportion of males to females when temperatures stay at about 25° C. (77° F.). The significance of this is not clear.

RESPONSES TO RAIN

An interesting association has been noticed between the breeding condition in many amphibians and the rainy season. In dry areas these amphibians depend on the onset of rain to stimulate their breeding impetus and the conditioning of their gonads, perhaps through increased food, plant and insect life, etc. There is also evidence that certain toads, especially in the desert areas of California, breed only in response to the presence of water— flash floods, spring rains, or whenever water arrives.

Just how water stimulates the breeding cycle is not always clear. In some cases it seems to be the very sensation of the rain on the body, but not in all. Experiments have shown that if the adrenal glands of one frog, *Rana cyanophylyctis*, are removed, it will not ovulate or spawn either in rainy or dry conditions.

Quite the opposite happens with the squirrel monkeys.[125] These animals live in a tropical area, and the length of day varies by only about eleven minutes throughout the year. So the length of the day does not affect the hormone release that leads to mating. Even though the young are born in the wet season, mating always takes place in the dry season, when the temperament of the males changes completely, so that they take over control of the groups, which at other times are ruled by the females. The males get thicker, fluffier coats, become fatter, more masculine in appearance, and are more vocal and much more aggressive when their testes become functional. Does the dryness of the air stimulate the gonadal hormones in this case?

THE EFFECTS OF CROWDING AND INADEQUATE FOOD SUPPLY

The influence of population density on breeding is probably quite complex. It is essential for the encouragement of breeding in some birds, but has a negative effect on most other birds and animals. In the latter a dense population makes itself felt in two ways, by reducing the food supplies below a minimum level and by causing excessive predation of eggs. In some fish (gudgeon are a good illustration), a favorable balance between population and food encourages an increase in the number of females, which ensures the maximum number of offspring. But when overcrowding occurs, the number of females dwindles, and males preponderate. This ensures a lower birth rate until the food situation is rectified.

Some sea bass and a few other fish pass through hermaphroditic stages, and this is used very advantageously to change the relative numbers of each sex in good or bad times. When feeding conditions are good, there are more females, as with the gudgeon, and when food is scarce, there are more males. In some fish species, such as the perch and flounder groups, the females mature more rapidly, but with less actual growth, when the food supply decreases. Smaller females produce fewer eggs, and in many species undernourished females often produce eggs that cannot be fertilized.

Two things are obvious. First, the gonads of most animals are subject to influences that increase their development at the time they need to function, usually in the spring, and any migration to breeding sites is in response to these same influences. Second, there are factors at work that help to control the number of each sex in many animals, according to whether conditions are adverse or favorable.

The importance of an adequate food supply for reproduction is illustrated by a very familiar bird. Pigeons are physiologically capable of breeding over a long period, but they do not necessarily do so. Although many male wood pigeons possess testes that are in full breeding condition in early spring and remain so to the end of summer or well into autumn, actual breeding occurs only when feeding conditions are suitable.

Many birds experience the arrest of the breeding cycle in both sexes in drought conditions, and even though ovulation usually starts immediately when rain falls, it must still be related to the available food supply. Birds of prey always produce more young in an abundant season, and some animals even remain in breeding condition much longer under these conditions.

It is easy to see that reproduction in the animal world is dependent on many things, each of which can be of greater or lesser importance to any single species. Whether an animal's readiness to breed is influenced by weather, environment, abundance of food, courtship ritual, the presence of large numbers of the species, solitude, signals from the opposite sex, or any other factor, it is so completely conditioned to these things that it cannot function normally if they vary beyond a certain very narrow degree of tolerance; and only a relatively few species have been able to relinquish a rigid courtship ritual.

Appendix

Bibliography

Amphibians

Noble, G. Kingsley. *The Biology of the Amphibia.* Dover Publications, Inc., New York, 1955.

Birds

Allen, Glover Morrill. *Birds and Their Attributes.* Dover Publications, Inc., New York, 1962.

Armstrong, Edward A. *Bird Display and Behaviour.* Dover Publications, Inc., New York, 1964.

Beebe, C. William. *The Bird.* Dover Publications, Inc., New York, 1965.

Chisholm, Alec H. *Bird Wonders of Australia.* Michigan State University Press, East Lansing, 1958.

Fish

Chute, Walter H. *Shedd Aquarium Guide.* Shedd Aquarium Society, Chicago, 1950.

Curtis, Brian. *The Life Story of the Fish.* Dover Publications, Inc., New York, 1961.

Grant, E. M. *Guide to Fishes.* Department of Harbours and Marine, Queensland, 1965.

Idyll, C. P. *Abyss.* Thomas Y. Crowell Co., New York, 1971.

Marshall, N. B. *The Life of Fishes.* Wiedenfeld and Nicolson, London, 1965.

Ommanney. F. D. *A Draught of Fishes.* Longmans, Green and Co. London, 1965.

Whitley, Gilbert P. *Marine Fishes of Australia.* Jacaranda Press, Brisbane, 1962.

————. *Freshwater Fishes of Australia.* Jacaranda Press, Brisbane, 1964.

Mammals

Allen, Glover Morrill. *Bats*. Dover Publications, Inc., New York, 1962.

Kummer, Hans. *Social Organization of Hamadryas Baboons*. University of Chicago Press, Chicago, 1968.

Morris, Desmond. *The Mammals*. Hodder and Stoughton, London, 1965.

Troughton, Ellis. *Furred Animals of Australia*. Angus and Robertson, Sydney, 1948.

Reptiles

Cogger, Harold. *Australian Reptiles in Colour*. A. H. & A. W. Reed, Wellington, N.Z., 1967.

Ditmars, Raymond L. *Snakes of the World*. The Macmillan Company, New York, 1966.

McPhee, D. R. *Snakes and Lizards of Australia*. Jacaranda Press, Brisbane, 1963.

General

Dobzhansky, Theodosius. *Genetics and the Origin of Species*. Columbia University Press, New York, 1951.

Freeman, W. H., and Bracegirdle, Brian. *Atlas of Embryology*. Heinemann Ltd., London, 1967.

Goodrich, Edwin S. *Studies on the Structure and Development of Vertebrates*. Dover Publications, Inc., New York, 1958.

Hamilton, W. J., Boyd, J. D., and Mossman, H. W. *Human Embryology*. W. Heffer and Sons Ltd., Cambridge, England, 1947.

McGill, Thomas E. (ed.). *Readings in Animal Behavior*. Holt, Rinehart, and Winston, New York, 1965.

Prince, J. H. *Animals in the Night*. Thomas Nelson Inc., Nashville, Tenn., 1971.

Steyaert, Jacqueline. *Encyclopédie en couleurs de l'anatomie animale*. Marabout Université, Gérard et Cie., Verviers, Belgium, 1967.

Storer, Tracy I. *General Zoology*. McGraw-Hill Book Co. Inc., New York, 1943.

Journals

Auk
Australian Journal of Zoology
Biological Abstracts
Bioscience
Condor
Emu
Ibis
Journal of Animal Ecology
Journal of Royal Zoological Society of New South Wales
National Geographic Magazine
Nature
Science

Glossary of scientific names of animals mentioned in the text

1 *Mugil cephalus*
2 *Gasterosteus aculeatus*
3 *Macropodus opercularis*
4 *Betta splendens*
5 *Haplochromes multicolor*
6 *Hippocampus* spp.
7 *Phyllopteryx foliatus*
8 *Rhodeus amarus*
9 *Hynobius* spp.
10 *Ambystoma* spp.
11 *Amblyrhynchus cristatus*
12 *Fringilla coelebs coelebs*
13 *Lobipes lobatus*
14 *Lophortyx* spp.
15 *Sericulus chrysocephalus*
16 *Chlamydera nuchalis*
17 *Xanunomelus aureus*
18 *Chlamydera maculata*
19 *Grus antigone antigone*
20 *Amblyornis* spp.
21 *Anas superciliosa*
22 *Anas castanea*
23 *Anas gibberifrons*
24 *Aythya australis*
25 *Biziura lobata*
26 *Centrocercus urophasianus*
27 *Tympanuchus* spp.
28 *Fregata* spp.
29 *Cephalopterus ornatus glabricollis*
30 *Struthic camelus*
31 *Procnias tricarunculatus*
32 *Artamus* spp.
33 *Casmerodius alba*
34 *Leucophoryx thula*
35 *Colymbus cristatus*
36 *Podiceps cristatus*
37 *Gavia stellata*
38 *Mergus serrator*
39 *Larus marinus*
40 *Sterna naethetus*
41 *Sterna bergii*
42 *Diomedea exulans*
43 *Diomedea immutabilis*
44 *Melopsittacus undulatus*
45 *Antilocapra americana*
46 *Zalophus californianus*
47 *Callorhinus ursinus*
48 *Phascolarctos cinereus*
49 *Castor fiber*
50 *Callicebus* spp.
51 *Alopex lagopus*
52 *Myoprocta pratti*
53 *Lama guanicoe*
54 *Balaenoptera s. musculus*
55 *Mustela vison*

56 *Martes zibellina*
57 *Tachyglossus* spp.
58 *Cricetus cricetus*
59 *Oryctolagus cuniculus*
60 *Lemur catta*
61 *Perodicticus potto*
62 *Theropithecus gelada*
63 *Hylobates syndactylus*
64 *Eucalia inconstans*
65 *Sterna paradisaea*
66 *Pluvialis dominica*
67 *Calidris fuscicollis*
68 *Numenius tahitiensis*
69 *Archilochus colubris*
70 *Calidris melanotos*
71 *Dolichonyx oryzivorus*
72 *Diomedia epomophora*
73 *Pygoscelis adeliae*
74 *Aptenodytes forsteri*
75 *Aptenodytes patagonica*
76 *Lasiurus borealis*
77 *Lasionycteris noctivagans*
78 *Alytes obstetricians*
79 *Pipa* spp.
80 *Varanus gouldii*
81 *Moloch horridus*
82 *Naja hannah*
83 *Crocodylus porosus*
84 *Uroaëtus audax*
85 *Ciconia* spp.
86 *Leipoa ocellata*
87 *Pygoscelis papua*
88 *Spheniscus demersus*
89 *Thalarctos maritimus*
90 *Suricata suricatta*

91 *Dasypus novemcinctus*
92 *Cuniculus paca*
93 *Nyctereutes procyonides*
94 *Eira barbara*
95 *Nasua nasua*
96 *Sylvilagua floridanus*
97 *Microcebus murinus*
98 *Daubentonia madagascariensis*
99 *Pongo pygmaeus*
100 *Leontideus rosalia*
101 *Bagre marina*
102 *Ameiurus melas*
103 *Larus atricilla*
104 *Puffinus puffinus*
105 *Branta canadensis*
106 *Haliaeetus leucocephalus*
107 *Pagophilus groenlandica*
108 *Mirounga leonina*
109 *Galago crassicaudatus*
110 *Atilax paludinosus*
111 *Lepus europaeus*
112 *Hydropotes inermis*
113 *Ursus horribilis*
114 *Pteropus polyocephalus*
115 *Macaca mulata*
116 *Cebuella pygmaea*
117 *Leontocebus* spp.
118 *Macaca sylvana*
119 *Propithecus verreauxi*
120 *Cavia porcellus*
121 *Coturnix corturnix japonica*
122 *Mugil curema*
123 *Cyprinus carpio*
124 *Melanogrammus aeglefinus*
125 *Saimiri sciureus*

Index